RELIABILITY AND RESILIENCE OF THE ELECTRICAL GRID FOR AND WITH NUCLEAR POWER PLANTS

The following States are Members of the International Atomic Energy Agency:

AFGHANISTAN
ALBANIA
ALGERIA
ANGOLA
ANTIGUA AND BARBUDA
ARGENTINA
ARMENIA
AUSTRALIA
AUSTRIA
AZERBAIJAN
BAHAMAS, THE
BAHRAIN
BANGLADESH
BARBADOS
BELARUS
BELGIUM
BELIZE
BENIN
BOLIVIA, PLURINATIONAL
    STATE OF
BOSNIA AND HERZEGOVINA
BOTSWANA
BRAZIL
BRUNEI DARUSSALAM
BULGARIA
BURKINA FASO
BURUNDI
CABO VERDE
CAMBODIA
CAMEROON
CANADA
CENTRAL AFRICAN
    REPUBLIC
CHAD
CHILE
CHINA
COLOMBIA
COMOROS
CONGO
COOK ISLANDS
COSTA RICA
CÔTE D'IVOIRE
CROATIA
CUBA
CYPRUS
CZECH REPUBLIC
DEMOCRATIC REPUBLIC
    OF THE CONGO
DENMARK
DJIBOUTI
DOMINICA
DOMINICAN REPUBLIC
ECUADOR
EGYPT
EL SALVADOR
ERITREA
ESTONIA
ESWATINI
ETHIOPIA
FIJI
FINLAND
FRANCE
GABON
GAMBIA, THE

GEORGIA
GERMANY
GHANA
GREECE
GRENADA
GUATEMALA
GUINEA
GUYANA
HAITI
HOLY SEE
HONDURAS
HUNGARY
ICELAND
INDIA
INDONESIA
IRAN, ISLAMIC REPUBLIC OF
IRAQ
IRELAND
ISRAEL
ITALY
JAMAICA
JAPAN
JORDAN
KAZAKHSTAN
KENYA
KOREA, REPUBLIC OF
KUWAIT
KYRGYZSTAN
LAO PEOPLE'S DEMOCRATIC
    REPUBLIC
LATVIA
LEBANON
LESOTHO
LIBERIA
LIBYA
LIECHTENSTEIN
LITHUANIA
LUXEMBOURG
MADAGASCAR
MALAWI
MALAYSIA
MALI
MALTA
MARSHALL ISLANDS
MAURITANIA
MAURITIUS
MEXICO
MONACO
MONGOLIA
MONTENEGRO
MOROCCO
MOZAMBIQUE
MYANMAR
NAMIBIA
NEPAL
NETHERLANDS,
    KINGDOM OF THE
NEW ZEALAND
NICARAGUA
NIGER
NIGERIA
NORTH MACEDONIA
NORWAY
OMAN

PAKISTAN
PALAU
PANAMA
PAPUA NEW GUINEA
PARAGUAY
PERU
PHILIPPINES
POLAND
PORTUGAL
QATAR
REPUBLIC OF MOLDOVA
ROMANIA
RUSSIAN FEDERATION
RWANDA
SAINT KITTS AND NEVIS
SAINT LUCIA
SAINT VINCENT AND
    THE GRENADINES
SAMOA
SAN MARINO
SAUDI ARABIA
SENEGAL
SERBIA
SEYCHELLES
SIERRA LEONE
SINGAPORE
SLOVAKIA
SLOVENIA
SOMALIA
SOUTH AFRICA
SPAIN
SRI LANKA
SUDAN
SWEDEN
SWITZERLAND
SYRIAN ARAB REPUBLIC
TAJIKISTAN
THAILAND
TOGO
TONGA
TRINIDAD AND TOBAGO
TUNISIA
TÜRKİYE
TURKMENISTAN
UGANDA
UKRAINE
UNITED ARAB EMIRATES
UNITED KINGDOM OF
    GREAT BRITAIN AND
    NORTHERN IRELAND
UNITED REPUBLIC OF TANZANIA
UNITED STATES OF AMERICA
URUGUAY
UZBEKISTAN
VANUATU
VENEZUELA, BOLIVARIAN
    REPUBLIC OF
VIET NAM
YEMEN
ZAMBIA
ZIMBABWE

The Agency's Statute was approved on 23 October 1956 by the Conference on the Statute of the IAEA held at United Nations Headquarters, New York; it entered into force on 29 July 1957. The Headquarters of the Agency are situated in Vienna. Its principal objective is "to accelerate and enlarge the contribution of atomic energy to peace, health and prosperity throughout the world".

IAEA NUCLEAR ENERGY SERIES No. NR-T-3.36

RELIABILITY AND RESILIENCE OF THE ELECTRICAL GRID FOR AND WITH NUCLEAR POWER PLANTS

INTERNATIONAL ATOMIC ENERGY AGENCY

VIENNA, 2025

COPYRIGHT NOTICE

© IAEA, 2025

Printed by the IAEA in Austria
December 2025
STI/PUB/2110
https://doi.org/10.61092/iaea.pm2g-2s99

IAEA Library Cataloguing in Publication Data

Names: International Atomic Energy Agency.
Title: Reliability and resilience of the electrical grid for and with nuclear power plants / International Atomic Energy Agency.
Description: Vienna : International Atomic Energy Agency, 2025. | Series: IAEA nuclear energy series, ISSN 1995-7807 ; no. NR-T-3.36 | Includes bibliographical references.
Identifiers: IAEAL 25-01749 | ISBN 978-92-0-106825-5 (paperback : alk. paper) | ISBN 978-92-0-106925-2 (pdf) | ISBN 978-92-0-107025-8 (epub)
Subjects: LCSH: Nuclear power plants — Power supply. | Nuclear power plants — Electric equipment. | Nuclear power plants — Design and construction.
Classification: UDC 621.039.5 | STI/PUB/2110

FOREWORD

The IAEA's statutory role is to "seek to accelerate and enlarge the contribution of atomic energy to peace, health and prosperity throughout the world". Among other functions, the IAEA is authorized to "foster the exchange of scientific and technical information on peaceful uses of atomic energy". One way this is achieved is through a range of technical publications including the IAEA Nuclear Energy Series.

The IAEA Nuclear Energy Series comprises publications designed to further the use of nuclear technologies in support of sustainable development, to advance nuclear science and technology, catalyse innovation and build capacity to support the existing and expanded use of nuclear power and nuclear science applications. The publications include information covering all policy, technological and management aspects of the definition and implementation of activities involving the peaceful use of nuclear technology. While the guidance provided in IAEA Nuclear Energy Series publications does not constitute Member States' consensus, it has undergone internal peer review and been made available to Member States for comment prior to publication.

The IAEA safety standards establish fundamental principles, requirements and recommendations to ensure nuclear safety and serve as a global reference for protecting people and the environment from harmful effects of ionizing radiation.

When IAEA Nuclear Energy Series publications address safety, it is ensured that the IAEA safety standards are referred to as the current boundary conditions for the application of nuclear technology.

A reliable electricity supply is an essential resource for nearly all aspects of modern life. The reliability of the electrical system depends in part on having enough dependable generating units to generate sufficient electricity to always meet the demand for electricity with a margin. Nuclear power generating units can contribute to this reliability by being a predictable and dependable source of baseload power and voltage control with high inertia. To meet the aim of reliable electricity supply and the goals for low carbon emission electricity generation, many Member States are considering the construction of nuclear power plants.

However, the interaction between a nuclear power generating unit and the national or regional electrical power system deserves careful attention. An unstable and unreliable electrical grid may increase the number of reactor trips and the probability and duration of loss of off-site power and station blackout events. Grid related events can also have a direct impact on the safety, operability and availability of a nuclear power plant. Conversely, the electrical grid can become adversely affected by an unexpected trip or disconnection of a nuclear power plant as large, abrupt changes in electricity generation can cause significant perturbations in the system. This could lead to grid system instability and failures, which in turn would affect the nuclear power plant. Such events impact the quality and availability of off-site power that is needed to ensure the removal of decay heat after a reactor trip and during the shutdown state.

It is also important for reliable grid operation that the design and operation of nuclear power generating units, their connection to the transmission system and the transmission system itself are properly coordinated and controlled. Consequently, programmes and procedures to exchange information between stakeholders need to be established so that mutually agreed upon reliability measures can be implemented and maintained.

In addition to the shared need for reliability, there is a shared need for the ability to recover in a safe and timely manner from unanticipated or extreme events that affect the electrical grid. The events following the earthquake and tsunami at the Fukushima Daiichi nuclear power plant in 2011 highlighted the need for the timely restoration of vital off-site power to the plant. However, keeping nuclear power generating units on-line during extreme natural events can enhance the grid system's resilience to recover from severe disruption.

This publication discusses the design and operation and the coordination and control requirements for a reliable and resilient electrical grid to support the safe and reliable operation of nuclear power plants. It also discusses nuclear power plant design and operation fundamentals to enhance the reliability and resilience of the electrical grid.

The IAEA expresses its appreciation for the generous contributions of many Member States and is grateful to all the contributors listed at the end of the publication. The IAEA officers responsible for this publication were Q. Yu and A.N. Kilic of the Division of Nuclear Power and A. Duchac of the Division of Nuclear Installation Safety.

1. INTRODUCTION

1.1. BACKGROUND

The establishment, operation and maintenance of nuclear power generation in a country depends on the development of energy infrastructures, policies and regulations, including the responsibilities and capabilities of the organizations involved and their interfaces. One of these infrastructures is the national or regional electrical grid[1], and one of these interfaces is the interface between a nuclear power plant and the electrical system.

Whether it is a Member State's first nuclear power plant or a new nuclear power plant is added to the existing nuclear power generation, the interface between each new nuclear power plant and the electrical power system requires attention, as the power system is important for the safe and efficient operation of nuclear power plants. A reliable and resilient electrical grid reduces the risk of reactor trips and the probability and duration of events of loss of off-site power (LOOP) and station blackout (SBO). During grid failures, the restoration of off-site power to the plant in a timely manner needs to be ensured by having an available electrical grid that will recover rapidly from such severe events.

Even if the availability of the off-site power is ensured according to the operational requirements, the quality of the off-site power can also have a direct impact on a nuclear power plant's safety, operability and availability, as it may result in service interruptions and damage to plant equipment.

The reliability of the electrical grid can also be affected directly by the operation of nuclear power plants, since they are typically the largest generating units in the system. A large, abrupt change in electricity generation, such as an unexpected trip or disconnection of a nuclear power plant, can cause significant disturbance to the electrical grid. This, in turn, could affect the availability or power quality of the off-site power supply to the nuclear power plant.

Therefore, for the safe and reliable operation of nuclear power plants, it is essential to maintain the reliability and resilience of the electrical grid so that it is compliant with the power quality requirements of the nuclear power plant.

To minimize or avoid adverse impacts from nuclear power plant events on the electrical grid, and vice versa, the following provisions need to be considered:

(a) The electrical grid needs to be designed and operated so that loss of electrical connections to the nuclear power plant is a rare event and the grid is resilient enough to recover from severe failures and resupply the nuclear power plant as a priority in a timely manner.
(b) The electrical grid also needs to be designed and operated so that reactor trips and other nuclear power plant events (e.g. a loss of coolant accident causing voltage degradation at the standby transformer) do not affect grid security and power quality.
(c) The design and operation of the nuclear power plant need to be compliant with the characteristics of the electrical grid to minimize or eliminate the impact of adverse events on the grid or the operation and safety of the plant.

The interaction of the electrical grid with the design and operation of various nuclear power plant technologies was investigated in the 1980s (see Refs [1, 2]). The safety requirements related to the interactions between the electrical grid and the plant are established in IAEA Safety Standards Series No. SSR-2/1 (Rev. 1), Safety of Nuclear Power Plants: Design [3].

---

[1] In this publication, the term 'electrical grid' is used to describe the public electrical system to which the nuclear power station is connected. The grid comprises a transmission system and distribution systems (see the Glossary at the end of this publication) to which power plants of all kinds are connected.

IAEA Safety Standards Series No. SSG-34, Design of Electrical Power Systems for Nuclear Power Plants [4], provides guidance for the design of nuclear power plants' electrical power systems, including the requirements related to availability, power quality and the stability of off-site power.

The nuclear safety significance of the electrical grid was highlighted by the IAEA–NEA operating experience report, the so-called Blue Book, for the period 2002–2005 [5].

IAEA Nuclear Energy Series No. NG-T-3.8, Electrical Grid Reliability and Interface with Nuclear Power Plants [6], aims to assist Member States that are considering the introduction of their first nuclear power plant to ensure that they analyse all interactions between the electrical grid and the nuclear power plant. That publication explains the need for the nuclear power plant to have a reliable interface with the electrical power system and describes how the reliability of the connection can be assessed and implemented. Since its publication, it has become important to readdress the questions of reliability and resilience, owing to the following events that have had an observed adverse trend:

(a) A study by the European Commission's Joint Research Centre [7] reported that events on the electrical system external to the nuclear power plant, such as the local substation and the grid system, contributed to nearly one third of LOOP and SBO events. The majority of the remaining (i.e. plant based) events were also due to the plant–grid interface that is under the control of the plant owner/operator, such as generators, transformers and busbars.

(b) The most recent periodic report of LOOP event analysis from a Member State [8] identified adverse trends in both overall LOOP frequency and LOOP durations.

References [7, 8] and other similar reports cover only LOOP and SBO events. However, operating experience has shown that nuclear power plants may also be sensitive to degraded electrical conditions such as abnormal voltage or frequency, slowly cleared faults or large phase unbalance. Examples of such events are discussed in Refs [8, 9–14]. In such grid related plant events, where the off-site power remains available but has degraded quality, the integrity and functioning of plant systems and components using this degraded off-site power can become adversely affected. For example, the events that occurred in Sweden (Forsmark-1 and Forsmark-3), the United States of America (Byron-2, twice) and the United Kingdom (Dungeness B) involved degraded power supplies that jeopardized electrical equipment essential to maintain stable and reliable operating conditions (see Ref. [7]). As discussed in Refs [8, 11–14], several events resulted in damage to equipment or in unstable and abnormal reactor operating conditions, which resulted in exceeding operational limits and conditions (OLCs; these OLCs are described in IAEA Safety Standards Series No. NS-G-2.2, Operational Limits and Conditions and Operating Procedures for Nuclear Power Plants [15]).

The significance and contribution of LOOP and SBO events to the overall risk to nuclear power plants were demonstrated by the accident at the Tokyo Electric Power Company's Fukushima Daiichi nuclear power plant in March 2011 (see details in Ref. [16]). Since nuclear power plants, unlike other generating units, require a power supply to the plant equipment to maintain reactor cooling after a shutdown, restoring off-site power in a timely manner after a LOOP event is important for nuclear safety, especially if the on-site emergency generators are not available. The lessons learned from the Fukushima Daiichi accident highlighted the importance of the availability and resilience of the plant and the equipment that provides off-site power, as well as the importance of plant and grid design, configuration, operation, programmes and procedures to ensure these. Following this event, nuclear regulatory organizations and nuclear power plant operators around the world reviewed the robustness and potential unavailability of off-site electrical supplies under severe hazard conditions such as those that affected the Fukushima Daiichi nuclear power plant.

Nuclear power plants are generally able to provide a dependable, predictable and continuous source of electricity generation, and thus contribute significantly to grid reliability. However, there have been recent changes in the energy policies of many Member States that may adversely affect the controllability and reliability of their grid systems. The following examples indicate a few of these changes:

— A move to a market system in power generation so that individual generating units may change output at short notice because of changes in market prices;
— A rapid change in generation from variable renewable energy sources with limited predictability, requiring more flexibility from dispatchable generating units;
— A corresponding reduction in generation from fossil fuel (e.g. coal, oil, gas) fired synchronous generators, causing changes in the inertia, voltage regulation and frequency response of the power system;
— An increase in distributed generation sources (e.g. solar photovoltaic systems, smaller wind turbines), whose power and energy are difficult for the grid control centre to predict;
— An increase in the number of high voltage direct current (HVDC) interconnectors between power systems, which are precisely controllable but provide no inertia to the system.

These factors lead to a system with less predictability of generation, lower inertia and a reduction in generation capacity that can be directly controlled by the grid system operator. These changes may lead to poorer control of frequency and voltage and may have an adverse impact on grid reliability. The developers and operators of nuclear power plants need to monitor any such changes and, if necessary, review the assumptions in plant safety analyses.

In light of these issues, the Resolution reached during General Conference 67 [17] states that the Conference:

"Recognizes the need to enhance the support for grid and nuclear power plant interfaces, grid reliability, and cooling water usage, and recommends that the Secretariat collaborate on these matters with Member States that have operating nuclear power plants".

1.2. OBJECTIVE

This publication aims to review the needs, challenges and solutions regarding interfacing the electrical grid with nuclear power plants in Member States safely and effectively. The publication intends to assist with establishing and sustaining a reliable and resilient electrical grid to support the safe and efficient operation of nuclear power plants, as well as the contribution of nuclear power plants to enhancing the reliability and resilience of the grid system.

The intended users of this publication are the organizations of the Member States that are involved in the decision making regarding grid infrastructure and reliable grid design and operation for and with nuclear power plants. The aspects listed in Section 1.3 are considered and discussed from the point of view of:

— Member States considering the construction of their first nuclear power unit(s) (newcomer countries);
— Member States that already have nuclear power plants in operation and are considering the addition of more nuclear power units (expanding countries);
— Member States that have existing nuclear power plants in operation and wish to review the performance of their grid systems (established countries).

Specifically, the following entities are foreseen as users:

— Nuclear power plant owners/operators;

— Grid system operators/transmission system operators[2], transmission system owners[3];
— Energy planners (at governmental or utility levels);
— Regulatory bodies (e.g. nuclear and grid/energy regulators);
— Technical support organizations.

Guidance and recommendations provided here in relation to identified good practices represent expert opinion but are not made on the basis of a consensus of all Member States.

1.3. SCOPE

This publication provides information on the interactions between the reliability and resilience of the electrical grid and the safe and efficient operation of nuclear power plants by investigating the following:

— The main physical, technical and administrative elements of the electrical grid that are relevant to the safe and reliable design and operation of nuclear power plants;
— The reliability and resilience characteristics of the electrical grid that are needed by a nuclear power plant;
— The nuclear power plant's contributions to and impacts on the reliability of the electrical grid.

All these aspects are considered and discussed on the basis of international operating experience of electrical grids, including their planning, maintenance and upgrades in support of nuclear power generation. The publication also provides relevant nuclear power plant design and operation fundamentals that contribute to the reliability and resilience of the electrical grid.

The electrical grid in a country (or region) comprises power stations and other electricity generation sources of all kinds, the transmission system, the distribution system and the grid control centre. The reliability of the electricity supply to consumers — such as private houses and small businesses connected to the distribution system at low and medium voltages (typically up to 35 kV) — depends very much on the design and operation of the distribution systems. However, nuclear power plants do not generally have connections to such networks, so the reliability of these networks is not particularly relevant to them.[4] Consequently, this publication does not discuss in detail the issues related to the reliability of distribution systems, but rather deals mainly with the operation of large nuclear power plants, the transmission system and the way in which the transmission system is planned and controlled. Specific attention is paid to the transmission system near the nuclear power plant and its interface with it, as events involving LOOP are usually due to faults in that area.

The discussions in this publication do not comprise a comprehensive list of all of the relevant needs, challenges and solutions, but rather provide key concepts that could be considered in the process, as a minimum, according to current operating experience and technical and administrative fundamentals. Further, this publication does not provide a detailed and prescriptive implementation procedure to achieve a reliable electrical grid operation interfacing with nuclear power plants. It is rather a descriptive guidance that addresses major technical and administrative topics and processes, as well as the important roles and responsibilities of stakeholders in this process.

---

[2] 'Grid system operator' is also referred to as 'transmission system operator' or 'independent system operator' in various Member States. See the Glossary for the definition used for the purpose of this publication.

[3] The transmission system operator might not be the transmission system owner. This is the case where the operator is an 'independent system operator'. In a Member State, both arrangements may exist. For example, in the United Kingdom, until 2018, the transmission system owners in Scotland were independent of the system operator, while in England and Wales the transmission system owner was also the system operator.

[4] However, in many Member States, there has been a rapid growth in the number of small sized electricity generation sources connected to distribution networks, which can have a significant effect on the way in which the grid system is controlled to maintain reliability and stability.

The main body of this publication is divided into eight sections, including the introduction (Section 1).

Section 2 discusses the definition and description of major physical and administrative aspects of the reliability and resilience of an electrical grid for (and with) nuclear power plants, while Sections 3 and 4 discuss the design and operational elements and concepts for the provision of system reliability and resilience, respectively.

With the key elements and goals for reliability, a set of assessments can be performed on the existing and planned grid system, as discussed in Section 5. Upon the identification of gaps, weaknesses and needs by such assessments, specific technical and design provisions can be considered and implemented. Key aspects of technical/design solutions, together with practised technical solutions for this identification, are provided in Sections 6 and 7.

Section 8 aims to raise awareness of changes and evolution of factors in the nuclear power and electrical grid domains, as well as climate change, that may have an adverse impact on the reliability and resilience of the electrical grid and its connections to the nuclear power plant, so that Member States can evaluate their own situations and take any necessary actions.

The annexes discuss operating experiences with major blackouts on high reliability grid systems and with large frequency excursions. They also provide two case studies. One concerns quality assurance in the operation and maintenance of substations and power plants that are important for electrical grid security and the other concerns market activities affecting grid planning.

2. CONTEXT OF RELIABILITY AND RESILIENCE OF ELECTRICAL GRID FOR AND WITH NUCLEAR POWER PLANTS

The design and operation of an electrical grid are similar with or without nuclear power generation being involved. However, there are additional requirements that need to be considered owing to the specific characteristics of a nuclear power plant.

The security and reliability of the electrical grid and the safe and reliable operation of a nuclear power plant are mutually dependent. For example, nuclear power plants are often the largest power output units and, therefore, they are significant for maintaining the stability of the grid by providing large inertia as well as active and reactive power reserves. On the other hand, a sudden disconnection of such a large electricity generation source will cause a considerable change in the system frequency, a large change in the power flows in some transmission circuits and probably a notable change in voltage nearby, posing a challenge to grid system control. Conversely, different electrical transients and phenomena in the electrical grid may propagate and affect the auxiliary systems of a nuclear power plant, causing deviations from normal plant operation.

This codependence can be managed by suitable cooperation between the organizations that operate the nuclear power plant and the electrical grid. There needs to be a variety of technical and commercial agreements between these organizations related to the design, performance, planning, operation and maintenance of the nuclear power plant and of the electrical grid.

The introduction of a new nuclear power generating unit into a country or region will require some modifications to the design and operation of the existing transmission system. These modifications may be substantial to ensure that the electrical grid is sufficiently reliable to meet the additional requirements associated with nuclear safety and the performance expectations of the operators of the plant and the grid. Furthermore, the planned operation and maintenance procedures for the electrical grid and the

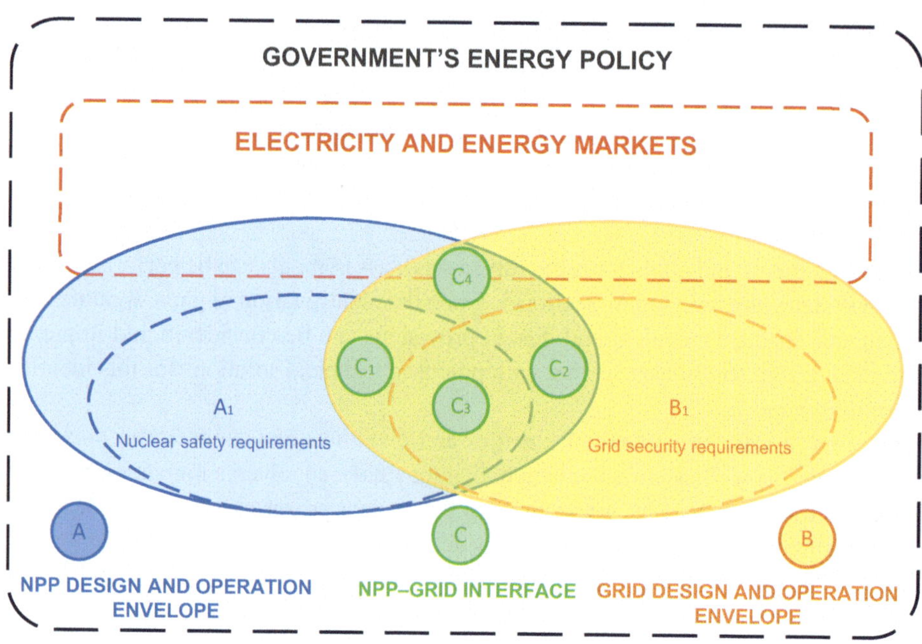

FIG. 1. Interfaces between a nuclear power plant (NPP) and the electrical grid (where $C = A \cap B$ and $C = C_1 \cup C_2 \cup C_3 \cup C_4$).

nuclear power plant need to be coordinated to avoid the disruption of system operation and to reduce risks for both sides.

Figure 1 shows the interfaces between a nuclear power plant and the electrical grid and summarizes the scope of this publication, which explores the experiences of Member States and the lessons learned concerning mutual reliability and resilience.

It is helpful to define and describe the domains shown in Fig. 1, which are discussed in more detail in Section 2.1, together with the terminology and concepts presented and discussed throughout this publication (some of which are also included in the Glossary at the end of this publication):

— Region A is the design and operation envelope of the nuclear power plant and covers all the equipment, activities and requirements used to meet the plant's safety requirements and performance goals, as further described in Section 2.2.1.
— Region A_1 is a subset of region A and is limited to those aspects of plant design and operation that are relevant to nuclear safety, as described in Section 2.2.3.
— Region B is the envelope that covers all equipment, activities and requirements of grid design and operation used for grid operation, as described in Section 2.2.2.
— Region B_1 is a subset of region B and is limited to those aspects of grid design and operation relevant to grid operational security, as described in Section 2.2.4.
— Region C includes the design and operation of equipment and the activities and requirements at the interface of the electrical grid and the nuclear power plant. This includes legal and regulatory matters related to the interface.
— Region C is further subdivided into regions C_1, C_2, C_3 and C_4:
  (i)   Region C_1 is not limited to the safety classified structures, systems and components (SSCs), but also includes equipment, activities and requirements, as described in IAEA Safety Standards Series No. SSG-30, Safety Classification of Structures, Systems and Components in Nuclear Power Plants [18], as follows:
  "Any SSC that does not contribute to any categorized function, but whose failure could adversely affect a categorized function (if this cannot be precluded by design), should

be classified appropriately in order to avoid an unacceptable impact from the failure of the function."

This type of SSC includes the standby transformer and the substation apparatus that connects to it, the voltage range at the connection point of the standby transformer and the arrangements for communication across the interface in emergency situations.

(ii) Region C_2 contains those elements that are important to maintaining grid security but are not necessarily related to nuclear safety, such as the voltage range at the connection point of the unit transformer, the systems to monitor and record the performance of the nuclear power plant and the agreements related to grid code compliance (see Glossary).

(iii) Region C_3 contains those elements that are related to both nuclear safety and grid security requirements, such as the coordination of protection systems across the interface, communication systems that inform the nuclear power plant operators of the status of the grid and the arrangements to restore supplies in the event of a blackout.

(iv) Region C_4 contains mainly the elements of the grid–nuclear power plant interface that are important for the availability of the nuclear power plant as seen from a grid perspective (for normal operation and during grid events caused by grid contingencies). This includes the main turbogenerator and its transformer, the turbogenerator controls for voltage and frequency, the nuclear power plant's electrical connections to the local substation and the various agreements related to the communication across the interface.

— 'Electricity and energy markets' (indicated in the diagram by a red dashed rectangle) are described in Section 2.2.5 and 'government's energy policy' (indicated in the diagram by a black dashed rectangle) is discussed as part of the role of the government in Section 2.3.8.

2.1. CONCEPTS OF GRID RELIABILITY AND RESILIENCE

Managing the reliability and resilience of the electrical grid is at the heart of the responsibilities entrusted to the grid system operator and the transmission system owner.

There are various definitions of electrical grid reliability and resilience in the literature and in the standards developed by industry experts. There are also instances of overlaps in these descriptions and interchangeable usage with other concepts, such as quality, availability, dependability and vulnerability. For the purpose of this publication, the concepts of electrical grid reliability and resilience are used as described in Sections 2.1.1 and 2.1.2, respectively.

2.1.1. Grid reliability

For an electrical grid, reliability is a measure of the ability of that system to deliver electricity to all points of consumption and receive electricity from all points of supply within accepted standards and in the desired amount (see Ref. [19]).

Currently, security and adequacy are necessary conditions to verify the classical property of 'reliability' for the electrical grid. Adequacy is the ability of the electrical grid to supply the load in all the steady states in which the electrical grid may exist, considering standard conditions. Security is the ability of the electrical grid to withstand sudden disturbances such as electric short circuits, unanticipated loss of system facilities or other rapid changes (e.g. in wind or solar generation).

Each Member State defines acceptable reliability differently, depending on many economic, geographical, political and social factors. However, a nuclear power plant requires a certain level of reliability, availability and power quality from the grid to reduce the likelihood of grid events that can affect the nuclear power plant adversely and to minimize the consequences of a grid related event on the nuclear power plant. This requires appropriate policies, plans and actions by all stakeholders regarding infrastructure development and operation.

In more specific terms, a reliable grid can be defined as a system where voltage and frequency are controlled within predefined limits and disconnections are infrequent events, as specified in more detail in Section 3.1.

2.1.2. Grid resilience

The International Council on Large Electric Systems (CIGRE) [20] defined resilience as *"the ability to limit the **extent, severity**, and **duration** of **system degradation** following an **extreme event**."* This definition is achieved through a set of key actionable measures to be taken before (anticipation and preparation), during (absorption) and after (sustainment of critical system operations, rapid recovery and adaptation) the event.

"The…definition clearly separates the definition of the properties from the ***key actionable measures*** that can be deployed [Before (**B**), During (**D**) and After (**A**) events] to achieve or enhance resilience.

- *The process of "**anticipation**" (**B**) refers to evaluating and/or monitoring the onset of foreseeable scenarios that could have disastrous outcomes.*
- *"**Preparation**" (**B**) is the process required by decision-makers to advance the knowledge gained during the anticipation phase from the resilience strategies to clear objectives to guide the deployment of measures.*
- *The process of "**absorption**" (**D**) is to meet defined objectives by means of which a system can absorb the impacts of extreme events and can minimise or avoid consequences.*
- *The "**sustainment of critical system operations**" (**D, A**) refers to the process of maintaining the operational capability of the impaired power system.*
- *The "**rapid recovery**" (**D, A**) process requires the operational response to contain or limit the consequence to the disruptive events.*
- *In the "**adaptation**" (**A**) process, changes are carried out in the power system management, defence and operational regimes, to contain and/or limit the undesirable situations."*

2.1.3. A reliable and resilient electrical grid for and with nuclear power plants

For the safety and reliability of operation of nuclear power plants, a reliable and resilient electrical grid has the following characteristics:

— A minimal frequency of grid related LOOP events (e.g. once in ten years or, ideally, never);
— A minimal occurrence of spurious reactor trips due to grid conditions;
— No restriction on the plant's operation in the safe and efficient generation of electricity;
— The capability and capacity to restore grid power supply to a nuclear power plant quickly after an interruption (e.g. within a few hours).

2.2. FRAMEWORK OF MUTUAL SAFETY, SECURITY AND RESILIENCE

The framework of mutual safety, security and resilience establishes the integrated relationship among the design and operation of the nuclear power plant, the performance of the electrical grid and nuclear safety requirements. This framework ensures that both the plant and the grid operate within their respective envelopes while maintaining compatibility and interdependence. Nuclear power plants need to comply with stringent safety requirements and operational commitments, while electrical grids need to guarantee reliability and security to support off-site power availability. Together, these elements form a coordinated system that safeguards people and the environment, ensures dependable electricity supply and enhances resilience against operational and external challenges.

2.2.1. Nuclear power plant design and operation envelope

Region A in Fig. 1 covers all activities and associated SSCs[5] and the human factors used in meeting the nuclear safety and design requirements and performance goals of the nuclear power plant. Also covered in this envelope are the associated programmes, processes and procedures that establish engineered and administrative controls for the design and operational activities, including the activities that involve the export and import of electricity.

The nuclear power plant design and operation envelope includes the aspects of design, operation and maintenance of the facility that are specific to meeting the requirements and expectations to:

— Comply with nuclear safety requirements (subsets C_1 and C_3 in Fig. 1), including the requirements for off-site power to be available at all times;
— Comply with the grid technical requirements applicable to the nuclear power plant (subsets C_2 and C_3 in Fig. 1) to ensure the export of electricity from the nuclear power plant to the grid without adverse impacts on the security of the electrical grid;
— Generate power as scheduled and coordinate plant outages and maintenance activities;
— Meet the commitments to and agreements with grid users and the grid stakeholders' network for dependability of the electrical grid.

2.2.2. Electrical grid design and operation envelope

Region B in Fig. 1 covers all those aspects related to grid planning, operation and maintenance to:

— Operate the electrical grid in compliance with grid security requirements (subset B_1 in Fig. 1, described in Section 2.2.4) for reliability;
— Ensure that electrical grid requirements are met and that they are compatible with nuclear safety and operational requirements (subsets C_1 and C_3 in Fig. 1), including the requirement for reliable provision of off-site power;
— Meet commitments made to electrical grid stakeholders and fulfil agreements with the electrical grid users for dependability;
— Foster economic performance and development of the electricity market to provide efficiency and economic benefits.

2.2.3. Nuclear safety requirements

Nuclear safety requirements (subset A_1 in Fig. 1) are an integrated and consistent set establishing the requirements that have to be met to ensure the protection of people and the environment, both now and in the future. Such protection is fulfilled by the provision of the following basic safety functions in all normal operations and in abnormal operations and accidents (see also Ref. [3]):

— Control of reactivity, that is, preventing uncontrolled reactor power increase and shutting the reactor down when required;
— Removal of heat from the reactor and from the fuel store;
— Confinement of radioactive material, shielding against radiation, control of planned radioactive releases and limitation of accidental radioactive releases.

The design and operation of a nuclear power plant in compliance with the nuclear safety requirements provide these basic safety functions by engineered features relying on the laws of physics and by

---

[5] Included in this envelope are technical, programmatic and human aspects for both the SSCs that are important (and related) to nuclear safety and the SSCs that are important (and related) to the plant performance of generating electricity.

engineered safety related SSCs. In addition, the nuclear power plant's management system supplements the basic safety functions by placing further administrative controls in the plant operation as described in IAEA Safety Standards Series No. SSR-2/2 (Rev. 1), Safety of Nuclear Power Plants: Commissioning and Operation [21]. These design and operational requirements are established by the laws and regulations of the country, and the nuclear power plant's compliance with them is reviewed, approved and overseen by the nuclear regulatory body, which authorizes and issues the operating licence for the plant. These are reflected in OLCs such as safety limits, safety system settings, limits and conditions for normal operation, and test and surveillance requirements [15]. They are also reflected in requirements for the design, operation and maintenance of safety related SSCs and associated programmes, processes and procedures that establish administrative controls for operational activities.

Many safety related SSCs require electrical power for their activation and control, so a reliable and available supply of electricity to them is important. The design of nuclear power plants requires the supply of electrical power either from off-site sources (i.e. from the external electrical grid) or on-site sources (i.e. from the main generator or from other internal generators, such as diesel or gas turbines). By applying the engineering concepts of diversity, redundancy, physical separation and functional independence in design and operation, the connection of the nuclear power plant to the off-site supplies improves the reliability of the supply of electrical power to safety related SSCs.

SSG-34 [4] provides recommendations and guidance on how to comply with the safety requirements for ensuring the reliability of the off-site supply in nuclear power plant design. Some of the key points are as follows:

(a) "The off-site power supply should have adequate capacity and capability to power plant loads in all modes of the nuclear power plant's operation" (para. 6.10 of SSG-34 [4]).

(b) "Off-site power should be supplied by two or more physically independent off-site supplies that are designed and located to minimize, to the extent practicable, the likelihood of their simultaneous failure" (para. 6.13 of SSG-34 [4]).

(c) "[A] single off-site power supply might be acceptable for reactors of a design that employs passive engineered safety features"[6] (para. 6.15 f SSG-34 [4]) but "Nuclear power plants with a single transmission line might have a forced outage rate that is higher owing to line tripping" (para. 6.16 of SSG-34 [4]).

(d) "As a minimum, each off-site power supply should have the capacity and capability to power all electrical loads required to mitigate the consequences of all design basis accidents and anticipated operational occurrences" (para. 6.17 of SSG-34 [4]).

(e) "Each off-site supply required for normal plant operation, startup and shutdown should have the additional capability to power all the normal electrical loads" (para. 6.18 in SSG-34 [4]).

The nuclear safety requirements related to the design and operation of the connection to the electrical grid depend on the reactor technology and on the design, operation and performance of the grid, which vary among Member States. These requirements may be explicitly defined in the nuclear power plant licensing basis or in the nuclear power plant design basis. They may also include requirements on the availability and the reliability of the electrical grid at the nuclear power plant connection point, such as the probability and frequency of occurrence of LOOP events or the likely future ranges of grid voltage and frequency.

---

[6] For example, some reactor designs employing passive safety systems without active safety related components may consider relying on a single transmission line, as described in SSG-34 [4]. However, the passive safety system concept and features differ in application and the acceptability of a single transmission line for each off-site power supply needs to be demonstrated by the safety and design analyses. These analyses should show that such an arrangement achieves the technical safety objectives as defined in SSR-2/1 (Rev. 1) [3], with the assurance of continuous and stable off-site power supply to all active systems.

2.2.4. Grid security requirements

The requirements for building up grid security (subset B_1 in Fig. 1) relate to the development, operation and maintenance of the electrical grid and are usually established and codified in operational standards or in publications regulating the technical relationship between the operators of the electrical grid and the grid users (e.g. grid codes).

The performance of certain components of a power plant (e.g. generator and its excitation system, turbine control system and auxiliaries, transformers) contributes to electrical grid security. The power plant operator's internal procedures and guidelines also contribute to this. Hence, the grid security requirements associated with those components and procedures need to be included in the nuclear power plant design, operation and maintenance requirements. For example, the capability of the generator to operate throughout a defined range of reactive power contributes to the need to control the voltage at the point of common coupling.

The special characteristics and operational requirements of nuclear power plants may be recognized in grid regulations or in agreements binding nuclear power plant operators and grid system operators.

2.2.5. Electricity markets

There are many differences among Member States regarding their commercial arrangements for the sale of electricity. At one extreme, the electricity industry is largely owned by the government, where the government sets the price for the sale of electricity generated by the power plants, and there is no wholesale electricity market. At the other extreme, in some Member States the power plants are owned by many private companies, the electricity industry has been completely deregulated, and the price of electricity is determined by market competition. There is a general trend worldwide towards deregulation and market competition in power generation.

The structure of the electricity market will affect the operation of power plants and of the grid. This is indicated in Fig. 1 by the overlap between electricity markets and the operation envelopes for the nuclear power plant and the grid. What is important for the reliability of the grid is that the structure of the electricity market not be in conflict with the reliability standards used by the grid system operator but to enable the grid system operator to comply with the reliability standards at all times.

From the point of view of a nuclear power plant, it has to be ensured that the electricity market rules are not in conflict with the plant's OLCs. It needs to be possible for the nuclear power plant to be operated in its preferred mode of steady full load for most of the time, and to plan and fix the dates for future refuelling and maintenance outages in the years ahead. The market arrangement would preferably not require nuclear power plants to make frequent or rapid changes in output.

Several types of market activity may have an impact on the electrical grid operation and its reliability, mainly related to the provision of reserve and the control of frequency. Typical market types are as follows:

(a) Long term market for energy. This refers to trading activities from years ahead to a few days or weeks ahead of real time. Power generating plant operators will adjust their plans for future operation and shutdown periods on the basis of the contracts that they have sold. The grid system operator will need to check that the long term plans do not lead to future periods of insufficient power generating plant capacity, which is unable to meet demand and provide adequate reserve.

(b) Short term market for energy. This refers to trading activities from a few days to a few hours ahead of real time. Trading is usually done in discrete time blocks, which might be blocks of four hours, one hour or even half an hour, depending on the Member State. This can lead to significant imbalances and hence to changes in the system frequency at the ends of trading blocks.

(c) Balancing market for energy. This refers to trading activities very close to real time, usually at the request of the grid system operator, to balance power generation with demand and control frequency.

If there is insufficient power generation capacity available for trading in these timescales, it will be difficult for the grid system operator to control the frequency adequately.

(d) Non-energy markets. In some Member States, there are payments made to power plants for the provision of ancillary services (e.g. operation in automatic frequency control mode, voltage control, the provision of reactive power, provision of reserve) and of capacity. Adequate provision of such services is very important for the reliable operation of the grid system, so market arrangements need to encourage the provision of these types of service.

A Member State that is considering changing the structure of its electricity market needs to carefully assess the effects that such a change may have on the ability of the grid system operator to control the system and maintain reliable operation, and any potential effect on the operation of a nuclear power plant.

2.2.6. The government's energy policy

As implied by Fig. 1, the development of the electrical grid and the electricity market is significantly affected by changes in government policy. The government's priority may be the lowest achievable price for electricity, or it may depend on targets for the reliability of the system. Government policy will also determine the extent to which energy generation sources and the electrical grid are owned and operated by competitive private companies or are owned by the state. The structure of the electricity market may be a direct result of government policy. In addition, the government may have targets for reducing pollution and stricter environmental limits.

Government policies and regulations will affect the planning and operational decisions by all stakeholders in electricity generation and transmission. The position of the government as a stakeholder in the development of the electrical grid is discussed further in Section 2.3.8.

2.3. KEY STAKEHOLDERS AND THEIR ROLES AND RESPONSIBILITIES

An electrical grid that is reliable and provides safe and secure electrical connections to nuclear power plants requires good cooperation between the key stakeholders. To achieve such good cooperation, it is important to understand the differing responsibilities and expectations of each of the stakeholders and the extent to which they influence the overall decisions and actions. The following sections summarize the responsibilities and expectations of the key stakeholders.

2.3.1. Grid system operator

The grid system operator is the company or organization that operates the grid control centre and is responsible for issuing instructions to power plants to increase or decrease energy generation to match supply with demand and to control frequency. The transmission system owner is the company or organization that owns and maintains part or all of the transmission system in the country. In some Member States, the grid system operator is the same organization as the transmission system owner. In other Member States, it is a separate organization and may be termed 'independent system operator' or 'independent transmission operator'. Reference [6] assumes that the grid system operator and the transmission system owner are the same organization, while this publication treats them as potentially separate organizations. The roles and responsibilities of the transmission system owner are summarized in Section 2.3.2.

The grid system operator will normally need to satisfy the requirements of the energy ministry or electricity regulator in the country, to achieve secure electricity supply within its control area at a minimum reasonable cost. This may include the requirement to plan and operate the system in accordance with published standards and procedures approved by the electricity regulator. The main responsibilities of this organization are operational planning and the real time operation of the transmission system in normal conditions, in disturbed conditions following anticipated faults and in emergency conditions. These include the following:

— Forecasting demand;
— Instructing power plants to change output where necessary to match demand and control frequency and to provide reserve;
— Ensuring that the system has sufficient redundancy and reserves to be secure against anticipated faults;
— Monitoring and controlling system frequency;
— Monitoring and controlling voltage at all points on the network;
— Monitoring and controlling power flows;
— Developing and practising procedures for the timely restoration of supply after local losses of supply or widespread blackouts and carrying out these procedures if such losses of supply or blackout occur;
— Communicating and collaborating with grid system operators in neighbouring countries or regions to which there are interconnections.

Activities for real time control are normally carried out at the grid control centre. Some of the necessary activities related to the loss of system integrity are described further in Section 4.3.5. The grid system operator may also be responsible for longer term actions, such as the following:

— Collaborating with the transmission system owner to plan outages on the transmission system for maintenance, testing and construction work;
— Longer term planning to verify that there will be adequate generating capacity to meet future demand and adequate requirements for reserve in the years ahead;
— Long term studies of the transmission system to understand the impact of changes in electricity demand, the addition of new power plants and the closure of old power plants;
— Long term studies of changes in power generation technology, such as the addition of large numbers of wind turbines and solar photovoltaic systems;
— Studies to establish the additions and reinforcements of the transmission system needed to accommodate new power plants and changes in demand;
— Analysing abnormal grid events to establish their causes and to reduce the likelihood and impact of similar events in the future;
— Reviewing and recommending changes to the published standards and procedures, such as the grid code, for the planning and operation of the grid system.

If a new nuclear power plant is to be connected to the system, the grid system operator will be responsible for the following:

— Agreeing with the NPP developer on the method by which the nuclear power plant is to be connected to the system and carrying out any electrical system studies to justify that method;
— Agreeing on the detailed design and protection arrangements for the substation(s) to which the nuclear power plant is to be connected, in collaboration with the transmission system owner and the nuclear power plant developer;
— Providing sufficient information to the nuclear power plant developer to allow a suitable assessment of the likely reliability of the proposed connection;
— Agreeing with the nuclear power plant developer on the technical performance criteria that the nuclear power plant needs to meet, as defined in the grid code or other documents.

2.3.2. Transmission system owner

As with the grid system operator, there may be a single transmission owner for the whole country, or there may be several, each with the responsibility for a particular geographical area or for particular circuits in the transmission system. The transmission system owner will normally have to satisfy the requirements of the energy ministry or electricity regulator in the country and the requirements of the grid system operator. The responsibilities of this organization will normally include the following:

— Planning outages of transmission system circuits and components for construction, testing, maintenance and repair, in collaboration with the grid system operator and the operators of power plants, and then carrying out such activities;
— Undertaking the detailed design and construction of additions and reinforcements to the transmission system, as identified by the grid system operator;
— Devising and implementing schemes for the electrical protection of the transmission system, which are coordinated with the protection schemes of transmission system users, in collaboration with the grid system operator and other users of the transmission system (including operators of nuclear power plants);
— Monitoring the performance of transmission system components to identify deteriorating performance and reliability, so that ageing components can be replaced in a timely manner to maintain system reliability;
— Investigating unplanned events and faults on the transmission system and carrying out emergency repairs when needed.

If a new nuclear power plant is to be connected to the system, then the grid system operator will be responsible for the following:

— Agreeing on the detailed ownership boundaries within the substation to which the nuclear power plant is connected and on the organization which will be responsible for the construction, future maintenance, etc., in collaboration with the nuclear power plant developer;
— Agreeing on the detailed design and protection arrangements for the substation(s) to which the nuclear power plant is to be connected, in collaboration with the grid system operator and the nuclear power plant developer;
— Constructing and commissioning any additions to the transmission system to allow the nuclear power plant to be connected.

2.3.3. Electricity regulator

The electricity regulator is the organization responsible for ensuring the proper functioning of the electrical grid to provide a reliable electricity supply to electricity consumers at the lowest reasonable cost. The nature of this regulator is different in different Member States. It might be responsible only for the electrical system as an 'electrical system regulator', or it might have wider responsibilities as an 'energy system regulator'. It may be part of the government (e.g. part of the energy ministry) or it may be a separate, independent organization, responsible to the government. In some Member States (e.g. the USA), there may be two or more electricity regulators with different areas of responsibility.

The electricity regulator is usually responsible for authorizing the technical and commercial rules for the operation of the electrical system, such as the grid code. Such rules may be different for different energy generation technologies and different unit sizes. The regulator may set targets for the control of the voltage and frequency and for the reliability of the supply to electricity consumers, and it may penalize the operators of power plants, transmission system owners or even grid system operators for actions or errors that challenge the reliability and security of the electrical system.

The electricity regulator may also be a 'market regulator' and be responsible for promoting market competition in the generation and supply of electricity and for controlling prices. It may provide information to the general public and inform the government on the performance of the electrical system.

Although the electricity regulator may have authorized a grid code or other technical rules, in some Member States the electricity regulator may give permission to individual generation sources to not comply with specific rules, where such non-compliance does not materially affect the reliability of the system and significantly reduces costs. The electricity regulator would not normally be directly involved in matters affecting nuclear safety. It is the responsibility of the developer or operator of a nuclear power plant to justify any change to the rules that they believe could adversely affect nuclear safety.

2.3.4. Nuclear power plant operator

The nuclear power plant operator is the company or organization that is authorized (licensed) to operate the nuclear power plant to generate electricity. The nuclear power plant operator may be the owner of the nuclear power plant, but there is a trend in many Member States for nuclear power plants to be partly owned by financial institutions who have no role in their operation.

The responsibilities of nuclear power plant operators are summarized in Ref. [22]. In particular, the nuclear power plant operator has the ultimate responsibility for the safe operation of the nuclear power plant, ensuring compliance with the plant's licensing basis (see IAEA Safety Standards Series No. SF-1, Fundamental Safety Principles [23], for further details), and for deciding the modes of operation and services that can be offered to the grid system operator. It is the main stakeholder of investment, purchase, construction, operation and maintenance of the plant, including establishing the interfaces and meeting commitments to other stakeholders. It is also the authority for preparation of preliminary safety, feasibility and business cases for the stakeholders, including establishing the grid requirements, as applicable at the decision making stage.

The nuclear power plant operator also has to comply with technical and operational requirements specified in the grid code or similar documents and with any contractual agreements that it has with the grid system operator, transmission system owner or other stakeholders. The nuclear power plant operator needs to ensure that the modes of operation and the ancillary services that the grid system operator requires are compliant with the plant's operating licence. If the country has a competitive market in the sale of electricity, then the nuclear power plant operator will also need to comply with the trading rules of that market.

If a new nuclear power plant is to be connected to the system, then the nuclear power plant developer/operator will be responsible for the following:

— Agreeing with the grid system operator and the transmission system owner on the method by which the nuclear power plant is to be connected to the system and on the design of the local substation;
— Understanding the characteristics of the grid at the point of connection and estimating the likely reliability of the connection, the frequency of LOOP events, etc., as an input to the safety case;
— Understanding all the grid technical requirements as specified in the grid code or similar documents, or as additionally specified by the grid system operator;
— Verifying that the proposed nuclear power plant design can meet the grid technical requirements and will be able to operate reliably for the expected full range of grid conditions, including abnormal events (as described in section 9 of Ref. [6]);
— Providing all necessary information to the nuclear regulatory body to establish the licensing basis of the nuclear power plant, including factors related to the reliability of off-site supplies, and to demonstrate that all planned operations will be within the licensing basis.

If there is a conflict between the grid code technical requirements and the requirements of the grid system operator, the electricity regulator and the nuclear regulatory body, it may be necessary for the nuclear power plant owner/developer to seek consent from the electricity regulator and the grid system operator to be non-compliant with specific requirements. Such consent would need to balance the effect of such non-compliance on the reliability of the grid system, the limitation of the licensing basis, and the cost and practicality of achieving compliance.

2.3.5. Nuclear regulatory body

The nuclear regulatory body is an authority, or a system of authorities, designated by the government of a Member State to have legal authority for conducting the regulatory process, including issuing authorizations, and thereby regulating nuclear, radiation, radioactive waste and transport safety [24]. It is generally a national entity, established and empowered by law, whose organization, management, functions, processes, responsibilities and competences are subject to the requirements of IAEA safety standards [24].

While the specific regulations governing the safe operation of a nuclear power plant vary among Member States — from very descriptive to very prescriptive — the general regulatory philosophy is consistent: the nuclear regulatory body requires that nuclear power plants are designed, built and operated within the bounds set by safety criteria for electricity generation. The nuclear regulatory body is a main stakeholder, as it reviews, assesses, oversees and authorizes the safe construction, operation and maintenance of the facility that is demonstrated within the nuclear power plant's design and licensing bases, to ensure public health and safety. It reviews the safety case (or safety analysis report) demonstrating the safe design and operation of the nuclear power plant, the integrated plant system's response and the ability of safety related SSCs to prevent, protect and mitigate the causes and consequences of anticipated operational occurrences and postulated design basis accidents. Also included in the safety case are the analyses and evaluation reports supporting the safe operation of the plant regarding a reliable supply of off-site power from the grid system and, if applicable, any services that the plant is expected or obliged to provide for secure and reliable grid operation.

As nuclear safety prevails over any other aspects, the nuclear regulatory body has a critical role in the decision making on plant design and operation and in the setting of any limitations on the support that the nuclear power plant can provide for reliable grid operations. The regulatory body is also the entity to eventually authorize the operation of a nuclear power plant, and the reliability of the electrical grid to which the nuclear power plant is connected is a factor in the decision for this authorization.

The particular activities of a regulatory body regarding grid reliability, therefore, are the following:

— Seeking assurances that the off-site power supply from the grid to the nuclear power plant is sufficiently reliable and that the likelihood of LOOP events and degraded off-site power conditions is sufficiently low;
— Understanding and authorizing the proposed nuclear power plant operational modes and schemes to support grid operation and ensuring that there are no adverse impacts on nuclear safety from providing the grid reliability support services.

2.3.6. Operators of other power plants

The operators of other power plants of all kinds, or of other generation sources such as wind turbines or solar panels, will be affected by the introduction of a new nuclear power generating unit into a country. The introduction of a large nuclear power unit operating at base load will affect the operation of other generating units, and — in a system with market competition — will affect market prices. It may be necessary for these other generation sources to provide more reserve and frequency control to ensure adequate control of frequency after a reactor trip to help maintain system security. There may be difficulties if this will require changes to the existing technical rules for generating plants or to any contractual agreements with those power plants. The nuclear power plant operator may also need to have agreements with the operators of other power plants connected close to the nuclear power plant.

In some Member States (e.g. France, Slovenia, the USA), other power generating units — such as hydroelectric or fossil fuel fired power generating units — are placed near the nuclear power plant. These additional generating units are identified as important for the availability of electrical power to the plant. They are used as the first backup to provide an alternative power source in case of a grid blackout or are used as an alternative or independent off-site power source in the plant safety case. The availability and reliability of such generating units and of the circuits connecting them to the nuclear power plant are important to the nuclear power plant. Consequently the nuclear power plant operator needs to have appropriate agreements in place with the operators of such plants.

2.3.7. Other network operators

As explained in Section 1.3, distribution networks do not generally have any direct effect on the reliability of electricity supplies to a nuclear power plant. However, for some nuclear power plants, the substation to which the nuclear power plant is connected may also provide connections from the transmission system to a local distribution network. Hence, there could be an interaction between the activities carried out by the distribution network operator on its equipment in the substation and the supplies to the nuclear power plant. For this reason, the nuclear power plant operator may need to enter into agreements with those network operators concerning work in the substation and any other activities that could interact with the grid connection for the nuclear power plant.

Where the transmission system of a country or region is connected to the network of another country or region, the introduction of a nuclear power plant in one network may require the grid system operator to negotiate changes to its agreements with the grid system operator in the other system.

2.3.8. Government

The national or federal government has the overall responsibility for the policy of developing a secure, reliable and resilient electrical system, in accordance with its energy plans for the country. The government may exercise this responsibility via one body (e.g. the energy ministry) or via several bodies within the government or under government control, depending on how the country chooses to organize. The authority may impose legislative and regulatory obligations and goals (as well as incentives or deterrents) directly on the grid system operators, the transmission system owners, and the owners and operators of power plants of all kinds. Alternatively, the government may impose some obligations indirectly by delegating responsibilities to other organizations, such as the electricity regulator. Whether imposed directly by the government or indirectly via an electricity regulator, such policies and regulations will affect the planning decisions made by all the stakeholders in the electrical system.

The government may develop a national position for additional power generation to meet the forecast rising demand, and this may include the introduction of a new nuclear power unit. It is important for the government to understand that investment in additional power generation will need to be accompanied by investment in strengthening the transmission system to maintain or improve the reliability and security of the electrical system.

During the entire policy making process, the government needs to collect feedback from all stakeholders to ensure that it understands the potential benefits and disadvantages for each party and for the system as a whole. Particularly for grid reliability and resilience, the government needs to consider the feedback from the grid system operator and electricity regulator related to matters such as links to neighbouring countries or networks, the age and reliability of the existing infrastructure, future changes in electricity demand, the benefits of alternative generation options and the possible need to change regulations or standards.

A robust government policy for the electrical system normally depends on reliable forecasts for the future electricity demand in the country. Such forecasts require cooperation between the energy ministry, the grid system operators, the transmission system owners, and the suppliers or utilities who supply electricity consumers, etc. The forecasts need to include a range of scenarios to take account of uncertainties and need to be updated regularly.

2.3.9. General public

Public acceptance is a very important issue for nuclear power, particularly for the introduction of new nuclear power plants [25]. Hence, it is important that the government and other stakeholders (such as the nuclear power plant developer) develop a policy to inform the general public of its plans and of the benefits of nuclear power for the country and the general public. This will help to foster the cooperation necessary to achieve general acceptance.

The development of a grid system able to provide reliable and secure electrical connection to a nuclear power plant may require the construction of new infrastructure such as new overhead lines and substations that are far from the location of the nuclear power plant. The experience in many Member States is that there is often strong public resistance to the construction of new overhead lines. Consequently, members of the public living far from a new nuclear power plant are potential stakeholders with an interest in the development. Hence, the government's efforts to obtain public acceptance have to concentrate not just on the public living near the new nuclear power plant, but also on people living farther away who may be indirectly affected.

2.4. MAIN ADMINISTRATIVE CONTROLS

The main administrative controls that assist the reliability of the grid system and the reliability of the connections to the nuclear power plant can be grouped into three categories:

(a)   Policies, rules and regulations;
(b)   Internal procedures within companies and organizations;
(c)   Agreements between companies and organizations.

2.4.1. Policies, rules and regulations

The government (or its energy agency or energy regulator) sets the national policy and associated regulations, and it is important that these policies and regulations have a secure and reliable grid system as one of their main objectives. The desire for the lowest reasonable cost electricity supply, and for achieving certain environmental targets, needs to be balanced with the need for a reliable system. As stated in Ref. [25], it is necessary for the government to evaluate the "State's long term energy needs and supply options, consistent with its social and economic development goals."

An important way to achieve a reliable grid system, as indicated in Sections 2.3.1 and 2.3.2, is to require the grid system operator and the transmission system owner to plan and operate the system in accordance with published standards and procedures approved by the electricity regulator or other government agency. Where these organizations are private companies and not government agencies, there can be financial incentives or penalties to encourage strict compliance with the standards or related to certain measures of performance, such as the control of system frequency and system voltage or the number and duration of disconnections of electricity consumers and power plants.

For power plants of all kinds, the main requirement will be to comply with technical regulations, such as those in a grid code or in similar documents, as described in Section 3.3.1. In most Member States, compliance with such requirements would be a precondition for being connected to the grid system. Such regulations usually specify the technical performance capabilities of the generating unit that need to be demonstrated when the unit is first connected and then maintained during the life of the plant. The regulations may also specify the information that is required to be exchanged with the grid system operator and the transmission system owner and require the power plant operator to comply promptly with instructions from the grid system operator. The government or electricity regulator may also impose rules to prevent certain commercial practices where these may disadvantage consumers or make it harder for the grid system operator to control the system.

2.4.2.   Internal procedures

The various stakeholders in the electrical system will need to have internal procedures to allow them to comply with the various regulations mentioned in Section 2.4.1.

The grid system operator and transmission system owner will need to have internal procedures to ensure that all design activities and operational activities that are necessary to comply with the security and other standards are carried out. They will also need to have internal procedures to ensure correct communication with the nuclear power plant and coordination of any work in its vicinity. These procedures for work related to a nuclear power plant may need to be different from the equivalent procedures for work related to other power plants, because of the need to consider the requirements of the licensing basis of the nuclear power plant. The nuclear power plant needs to clearly communicate any such special needs.

The nuclear power plant operator will have many internal procedures needed for normal operation of the plant in accordance with its licensing basis. These procedures will control activities for the following:

— Normal system operations, such as turbine, generator, unit transformer operations and other plant electrical power systems interfacing with the grid;
— Normal plant evolution operation, such as startup, partial or full power generation and plant shutdown;
— Alarm response, such as control room electrical panel alarm response for grid and generator voltage and frequency alarms;
— Abnormal operation, such as complete or partial LOOP, act of nature, generator voltage disturbances, electrical grid disturbances, main turbogenerator trip, loss of load and load rejection;
— Checking and, if necessary, rejecting any instruction from the grid system operator if it goes outside agreed parameters or outside the licensing basis.

It is helpful to grid reliability and security if the nuclear power plant's procedures for abnormal grid conditions (e.g. voltage or frequency outside normal range) avoid tripping the reactor as far as possible.

Most nuclear power plants have a plant simulator for training control room staff. It is advisable that the training procedures include simulation of various grid events, such as a period of abnormal voltage, a period of abnormal frequency, loss of connection to the unit transformer, loss of connection to the standby transformer or a complete LOOP.

2.4.3.   Agreements between companies and organizations

The experience in Member States is that it is necessary to have legally binding agreements between the stakeholders in the electrical grid. This section mainly considers the agreements between the grid system operator, the transmission system owner and power plant operators.

The structure of these agreements depends on the structure of the electrical power system and the way that it is regulated, and it is specific for each country. Generally, the agreements between the grid system operator, transmission system owner and power plant operator can be categorized as follows:

(a)   Grid usage. This category covers agreements needed by the grid system operator and power plant operators to run their business (e.g. grid connection, power take off or injection, communication, planning and scheduling).
(b)   Grid services. This category covers agreements needed by grid and market operators to operate electrical grids and power markets (e.g. ancillary services, black start, reserves, auction mechanisms for allocation of interconnection capacity).

The agreement for grid usage is typically called a 'connection agreement', and generally requires that the power plant complies with the technical requirements in the grid code or similar technical documents. The connection agreement will generally specify the details of the grid connection, the timescales for construction (if it is a new connection) and, most importantly, the capacity of the connection (the maximum permitted export and import power). It is important that this agreement, or other documents to which it refers, clearly specifies the ownership boundary and the responsibility for safety rules and for the operation and maintenance of all plant and equipment at the interface. The communication between power plant operators and the grid system operator and transmission system owner, for planning and operation, also need to be specified clearly, either in this connection agreement or in other documents to which it refers.

The agreements for grid services will typically relate to the provision of ancillary services (such as frequency control, reactive power generation or compensation, inertia or simulated inertia) and any arrangements for payment for these services. The agreements may also include provision for participation in a market mechanism for balancing generation and demand in real time or a market mechanism to provide reserve. Some power plants (although almost certainly not nuclear ones) may have an agreement to provide black start capability (as defined in the Glossary).

The operators of the nuclear power plant may need additional agreements with the grid system operator to allow the nuclear power plant operator to comply with its nuclear licensing basis. These agreements may include the following provisions:

— The grid system operator and/or the transmission system owner to notify and subsequently agree on any proposed changes to the transmission system or its components close to the nuclear power plant point of connection, so that the nuclear power plant operator can correctly assess the effect of the changes on the reliability of the connection or any other requirements of the nuclear licensing basis;
— The grid system operator and/or the transmission system owner to notify and agree on any planned outages on defined circuits and components close to the nuclear power plant point of connection (this may be more information than is normally provided to other power plants);
— The grid system operator and/or the transmission system owner to produce a formal report for the nuclear power plant operator, on request, describing the performance of transmission circuits and substations relevant to the nuclear power plant's secure grid connection;
— The grid system operator and/or the transmission system owner to ensure that restoration of the grid connection to the nuclear power plant, following an event of loss of connection, is given a high priority.

The nuclear power plant operator has the right to refuse an instruction from the grid system operator, or to shut down in certain circumstances, if that is necessary to avoid breaching the nuclear licensing basis.

The grid system operator may also have other agreements to assist in controlling the system, such as agreements with some large electricity consumers to allow for disconnection of power for those consumers in case of certain grid system emergency conditions or as a commercial service in non-emergency conditions.

The nuclear power plant operator may have separate agreements with other power generating plants located near the nuclear power plant, or with distribution network operators who have a connection to the transmission system close to the nuclear power plant, to ensure good communication of any maintenance or other work that may interact with the nuclear power plant.

2.5. THE CONTRIBUTION OF NUCLEAR POWER PLANTS TO GRID RELIABILITY AND RESILIENCE

Nuclear power plants can contribute significantly to the reliability and resilience of an electrical grid, in the ways summarized in Table 1.

TABLE 1. THE CONTRIBUTION OF NUCLEAR POWER PLANTS TO THE RELIABILITY AND RESILIENCE OF THE GRID

Grid needs	NPP contribution for grid reliability	NPP contribution for grid security
System balancing and frequency control	— Continued operation for normal range of frequencies — Significant inertia — Ability to load follow — Large load frequency control capability	— Continued operation during large frequency deviations — Small power reduction when the frequency is falling
Voltage control	— Continued operation for normal range of voltages — Wide range of reactive power control	— Continued operation during abnormal voltage events
Transient stability	— Large voltage and power damping capacity — Fault ride-through capability	— Large current contribution to grid short circuits
Control in emergency situations	n.a.	— Ability to perform load rejection (power cutback) to avoid circuit overloads — Ability to trip to house load operation if disconnected — Ability to control frequency and voltage in island mode operation if disconnected with demand
Fuel security	— Sufficient fuel loaded in the reactor or in store for several years of operation	

Note: n.a.: not applicable; NPP: nuclear power plant.

As nuclear power plants use synchronous generators (turbogenerators), their generators have similar characteristics to the turbogenerators used at other thermal power plants, such as those fuelled by coal, oil or gas, although generally with a larger unit size. Nuclear power plants are also generally built more robustly than these other power plants. They can contribute to grid reliability, provided that they have been designed to be compatible with grid requirements and have the following attributes:

— The ability to operate continuously at steady load (typically full load) throughout the defined normal range of frequency and voltage;
— The ability to continue to operate during and after transient disturbances on the grid system, such as the voltage depression caused by a nearby fault due to a lightning strike ('fault ride-through capability', as defined in the Glossary);
— The ability to generate or absorb reactive power and to change reactive power dynamically, as required to control system voltage;
— The ability to provide damping of system oscillations using its excitation control system;
— A known high level of inertia;
— A known high level of short circuit infeed to the grid.

Currently, the majority of nuclear power units in the world are operated at steady base load. However, depending on their design, many nuclear power units could operate flexibly, changing generated output when needed, to assist in balancing generation with demand to help control of system frequency and power flows. This is discussed in detail in IAEA Nuclear Energy Series No. NP-T-3.23, Non-Baseload Operation in Nuclear Power Plants: Load Following and Frequency Control Modes of Flexible Operation [26]. While nuclear power units are unlikely to be able to achieve the rapid response achieved by hydroelectric generators and thermal power plants that have been specially optimized for a fast response, many nuclear power units, particularly new nuclear power units, will be able to provide the following:

— Load following;
— Load frequency control;
— Rapid power cutback (load shedding).

In addition, depending on the design, a nuclear power plant may be able to contribute to the resilience of the grid system if it has the following attributes:

— The ability to continue to operate, perhaps at slightly reduced load and for a limited period, for a wider range of frequency and voltage (outside the normal operating range) during major system disturbances;
— The ability to continue to operate, supplying to some local demand, if the plant and the local demand have become separated from the rest of the grid system ('island mode operation', as defined in the Glossary);
— The ability to avoid tripping the reactor if the nuclear power plant is disconnected from the grid ('house load operation', as defined in the Glossary).

Nuclear power plants have high capital cost but low running costs (fuel costs) and generally take a long time to restart if a reactor trips unexpectedly. Hence, there can be a strong financial incentive for the operators of a nuclear power plant to have and maintain the capability to continue operation for a wide range of voltage and frequency during and after transient events, and if separated from the grid with or without local demand. However, not all reactor designs are capable of island mode operation or house load operation because of the need for a very rapid, controlled load reduction before such operation.

The generators in many nuclear power plants have a high inertia constant so that they provide more inertia to the system than other thermal power plants with similar generating capacity. This is becoming more important with the increase in many Member States of power generation sources that provide little or no inertia to the system. Similarly, there is an increase of generation sources that provide little short circuit infeed to the grid, so the infeed provided by synchronous generators, such as nuclear power units, is important to maintain fault levels. This is to allow the continued correct operation of protection systems that detect overcurrent and of the inverters used by HVDC converters and in many wind turbines and solar photovoltaic systems.

Nuclear power plants need to be shut down periodically for refuelling or to carry out inspections, tests and maintenance as required by the nuclear regulatory body and in accordance with good engineering practices. Because of the constraints of refuelling and the licensing by the nuclear regulatory body, the dates for planned outages are relatively fixed and can be planned and notified to the grid system operator and transmission system owner many years in advance. Between the planned outages, nuclear power plants are almost always operated at steady full load. This is in contrast with other forms of power generation that can and do change their operating regime and planned outage dates, sometimes at short notice, because of market conditions. It is also in contrast to some forms of renewable energy generation, which change output constantly and whose output cannot be predicted with confidence more than a few days in advance. The better long term predictability of the output from nuclear power units allows better long term planning for system reliability by the grid system operator and for transmission circuit outages by the transmission system owner.

Most nuclear power plants currently operate on an 18 month fuel cycle, with approximately one third of the fuel replaced at each outage. Plants that operate on a shorter or longer fuel cycle or that carry out refuelling on load replace their fuel at a similar rate. It is feasible for a nuclear power plant to store the fuel needed for the next refuelling on site. This means that most nuclear power units can typically continue to operate at full load for many months even if there is an interruption in the delivery of new fuel to the plant. This is in contrast to power plants using fossil fuel. Plants burning coal or oil generally cannot store the fuel for operation on site for more than a few months (in the case of coal) or a few weeks (in the case of oil). Plants that use piped natural gas could lose fuel supply within a few hours because of interruptions on the gas network. Similarly, a prolonged period of dry weather can greatly reduce the 'fuel' available to hydroelectric power plants. By comparison, nuclear power units are relatively immune to interruptions of fuel supply and, hence, assist grid security.

3. DESIGN AND OPERATIONAL ELEMENTS OF A RELIABLE ELECTRICAL GRID FOR AND WITH NUCLEAR POWER PLANTS

The meaning of reliability for the electrical grid in a country is summarized in Section 2.1.1. This section provides more detail on the concept of reliability.

3.1. CHARACTERISTICS OF A RELIABLE GRID

As described in Ref. [6], the desired performance characteristics of an electrical grid that are relevant to a nuclear power generating unit include the following:

— Frequency is well controlled, for example within ±1% of the nominal frequency for the majority of the time. Frequency may go outside such a nominal range, but only for short periods and on a few occasions per year. This may be allowed, for example up to +3/−5%. However, some designs of nuclear power plant may require a narrower range of frequency variation.
— The voltage at the substation(s) to which the nuclear power plant is connected is well controlled, for example within ±5% of the nominal value on the transmission system for the majority of the time. The voltage can go outside this range for short periods on a few occasions each year, for example, up to 10% deviation may be allowed depending on the nominal voltage.
— The grid is able to continue normal operation after most single unplanned events. This will require operation to an $N-1$ criterion or better, as described in Section 4.1 and in the Glossary.
— Events that disconnect parts of the grid or lead to blackout of a major part of the grid are rare (e.g. significantly fewer than one per year), particularly in that part of the grid to which the nuclear power plant is connected.
— In the case of a regional blackout, power for essential services, including off-site power for the nuclear power plant, can be restored in a few hours.

It may be possible to upgrade the grid to meet the requirements of a particular nuclear power plant design. Alternatively, it may be possible to modify the nuclear power plant design to be compatible with the characteristics of an existing grid. These options need to be discussed between the grid system operator, the company planning to develop the nuclear power plant and the designer of the nuclear power plant at an early stage. It is important that the nuclear power plant designers are informed in detail about the existing performance of the grid and any planned improvements. Conversely, the grid system operator needs to be informed about any changes in reliability requirements of the nuclear power plant.

Two international specifications exist for nuclear power plants for operation in a typical, reliable electrical grid with the characteristics summarized above (see Refs [27, 28] for further details). The same topic is also covered in Section 3.3.2. It is likely that most reactor vendors will ensure that, as a minimum, their designs are compliant with these specifications.

3.2. GRID SYSTEM PERFORMANCE FOR NUCLEAR POWER PLANTS

From the perspective of the grid system operator, the most important grid performance measures during normal and anticipated conditions are typically related to frequency and voltage deviations and major grid disruptions. Many aspects and factors influence the management and control of deviations within the acceptable ranges during steady state and dynamic conditions, such as the following:

— Loading of different network facilities and/or equipment (particularly the transmission lines, including tie lines and power transformers);
— Availability of important network facilities/equipment and generators connected to the grid;
— Efficiency of the installed protection and control systems;
— Frequency and duration of grid blackouts and/or formation of system islands;
— Unsupplied electricity due to grid limitations;
— Frequency of cross-border congestions, availability of ancillary services and cross-border deviations compared with the planned values.

Specific to an existing or new nuclear power plant, the grid performance depends on the control and management of frequency and voltage variations at the grid connection point, which is mostly influenced by the design and the operational characteristics of the following systems and equipment:

(a) Systems and components converting primary energy into electricity (e.g. the steam generator, turbine, generator);
(b) The substation(s) to which the nuclear power plant is connected (e.g. circuit arrangements, protection systems);
(c) Transmission circuits connecting the electrical grid to the substation(s) to which the nuclear power plant is connected.

As for any power generating unit, the nuclear power plant's active systems and components (e.g. motors, pumps, inverters, battery chargers, valve actuators) require definite levels and ranges for voltage, frequency and short circuit power. A good electrical grid has the following performance attributes:

— It maintains system frequency and voltage at the connection point as stable as possible with deviations being rare and/or not significant enough to jeopardize the proper functioning of plant equipment and safety systems. Further attributes of this feature are:
  - The system frequency is kept close to the nominal value. Deviations from the nominal value need to be minimized and the largest instantaneous deviations upwards or downwards cannot reach the maximum/minimum allowed value.
  - Voltage at the connection point is kept within the permitted range between a maximum and a minimum value, avoiding extreme voltage dips or spikes and with satisfactory symmetry of the sinusoidal wave.
  - During nuclear power plant upset conditions (e.g. reactor trip, loss of off-site power, pipe break) the voltage remains within a permitted range between a maximum and minimum value.
  - The rate of change of frequency (ROCOF) during deviations is mitigated.
— It removes symmetrical and asymmetrical electrical faults on the electrical grid without disconnecting the nuclear power plant and by keeping stable operation after the fault clearances.

— It delivers a power quality compliant with applicable International Electrotechnical Commission (IEC) or other international standards.

3.2.1. Reliable and stable frequency

Maintaining a reliable and stable grid frequency is essential for the secure and efficient operation of both the electrical system and nuclear power plants. Frequency control involves coordinated planning and real time adjustments to balance generation and demand, ensuring system stability under normal and abnormal conditions. For nuclear power plants, frequency deviations can affect critical equipment and safety margins, potentially triggering protective actions or reactor trips. Conversely, an unplanned trip of a nuclear power unit can significantly impact grid frequency owing to its large generation capacity. Therefore, robust frequency management measures and clear operational provisions are vital to minimize risks and maintain overall system resilience.

3.2.1.1. Reliability measures for a stable frequency

The control of grid frequency is achieved by two main processes:

— Successive and rolling phases for energy generation and grid planning (e.g. years, seasons, weeks, a day or multiple days ahead, up to a few hours) to elaborate a generation schedule matching the total expected demand;[7]
— Real time control of the frequency, mainly achieved by adjusting the generation and/or demand once a frequency deviation is detected.

Concerning the nuclear power plant, the main means to control the frequency are as follows:

— Adjusting the active power output either automatically or manually;
— Performing load and frequency control with different response times (typically varying from a few seconds to several minutes) using several layers of control schemes with combined automatic and manual activation.

Member States have different provisions related to market arrangements (ancillary services) and controls that will help to regulate frequency. In some countries, these services are considered as mandatory and certain generating units are obliged to provide them, while in some countries there are fully functioning competitive markets with payment for the provision of these services. However, the electricity market arrangements and controls have no direct impact on frequency reliability and stability, because the implementation of the market positions is integrated in the generation planning by the grid system operators. However, delayed or anticipated changes to power generation or non-conformance of market participants to ramping requirements may create generation–demand imbalances, and therefore, resulting frequency deviations.

3.2.1.2. Impact of frequency management on nuclear power plant operation

As discussed in detail in Ref. [6], maintaining system frequency within defined limits is important for the proper operation of plant equipment and systems, particularly safety related equipment.

Large variations of system frequency may jeopardize electrical equipment at the nuclear power plant, in terms of maintaining stable and reliable operating conditions. For example, when the frequency decreases or increases, it may change the reactor coolant pump speed in a pressurized water reactor,

---

[7] The needs of the power buyers, sellers, market operators and grid system operators are integrated during these planning phases.

which then affects the coolant flow through the core and the core power distribution, together with the temperature variation along the core. This may cause the power distribution limits in the reactor core to be reached, resulting in automatic setting off of reactor power limitations or trip, which in turn may cause a large change in frequency (see Ref. [26]).

The safety case is elaborated considering a particular frequency range; the resulting safety margins are calculated using this range with impacts on reactor rod fall time, on the initial conditions used in the analysis of departure from nucleate boiling accidents and on the behaviour of the active components of the nuclear island. Therefore, if the hypotheses concerning the frequency behaviour used for the safety case are changing, then the safety regulator may ask to review the safety case accordingly.

3.2.1.3. Impact of nuclear power plant trip on grid frequency

As the nuclear power plant is likely to be one of the largest power generating units on the system, an unplanned trip of the plant will cause a substantial fall in system frequency. Furthermore, the disconnection of a nuclear power plant from the main transmission system (i.e. the plant becoming 'islanded') would challenge not only the nuclear power plant's on-line electrical power system loads in terms of high or low frequency and rapid ROCOF, but also the grid system loads. This impact on the electrical grid could occur even where a nuclear power plant is normally connected to a large and strong grid network with a stable frequency (see Ref. [6]).

3.2.2. Reliable and stable voltage

3.2.2.1. Reliability measures for a stable voltage

Control of the voltage is performed by the grid system operator at the grid substation levels and is achieved by two main processes:

— Periodic grid and generation planning (e.g. years, seasons, weeks, up to days ahead) to elaborate a map of the expected voltage at the grid substations compatible with grid operation criteria for voltage ranges and short circuit levels;
— Real time control of voltage levels at grid substations, mainly achieved by adjusting the reactive power generated or absorbed by the generating units of grid specific equipment (e.g. a static volt-ampere reactive compensator) as well as by rerouting some power transits.

Concerning the nuclear power plant, the main means to control the voltage are as follows:

— Adjusting the level of reactive power (generated or absorbed) at the generator terminals either automatically or manually;
— Controlling and stabilizing, by the generator excitation system, of the electrical parameters at the generator terminals (e.g. voltage, angle, power).

Nuclear power plants have an important role in stabilizing the voltage at the regional level because, in some Member States, owing to the large reactive power capacity, nuclear power plant substations are used by grid system operators as pilot points in secondary voltage control systems for defined grid areas.

3.2.2.2. Impact of grid operation on nuclear power plant operation

A degraded voltage condition on the transmission system to which the nuclear power plant is connected is an important contributor to LOOP events (see Ref. [29]). This demonstrates that good control of the voltage and the provision of sufficient reserves of reactive power are important for the safe and reliable operation of a nuclear power plant.

It is important for the nuclear power plant that the grid voltage is also controlled within the acceptable range to provide power to the auxiliary equipment when the reactor is shut down. If a nuclear power unit trips at a time when it is exporting a large amount of reactive power to the system, then the local grid voltage may fall after the trip. The voltage control arrangements on the grid have to ensure that the grid voltage remains within the acceptable range, specifically for the nuclear power plant SSCs,[8] after such an event (see Ref. [6]).

There is also a potential nuclear safety problem for the nuclear power generating unit if there is a disturbance that results in islanding of the nuclear power plant at a time when it is exporting a large amount of reactive power. Such a transient can cause abnormally high voltage locally, which could affect safety systems (see Ref. [6] for further details). Furthermore, a design basis accident at the nuclear power plant may cause the voltage to fall because of the additional loading in the plant response to the event. This needs to be considered in the plant design.

3.2.2.3. *Impact of nuclear power plant voltage control on grid operation*

Owing to the large capacity of the generators at most nuclear power plants, they act as important stabilizers against voltage and angle deviations on the grid (see SSG-34 [4]).

However, during operation, it is important to ensure that voltage control functions and associated parameters are compliant and communicated to the grid system operator. For example, the implementation of an inadequate reactive power limitation can be discovered only during a voltage fall event, accelerating the voltage fall and leading to a plant trip and probably to a regional voltage collapse.

3.3. TECHNICAL REQUIREMENTS FOR GRID RELIABILITY

Nuclear power plants need to comply with the technical requirements of the grid system operator to become a part of the electrical grid. These technical requirements bear the brunt of the planning, design and operation characteristics of generating units and their connection to the grid. The contents of and the enforcement process for the requirements differ among the Member States, depending on the regulatory framework, applicable national and regional industry standards, market arrangements and, more importantly, grid infrastructure and characteristics.

Some requirements are generic and are the same for all types of energy generation,[9] or they can be specific to the generation type accounting for the inherent technology (e.g. synchronous generators versus non-synchronous generators such as wind farms, nuclear versus coal fired boilers).

3.3.1. Requirements established by electricity regulators and grid system operators (grid codes)

Many Member States have established sets of rules, often called grid codes, that are specified or approved by a government agency or the energy regulator of the country. Compliance with the grid code may be enforced by the grid system operator or another regulatory body.

The contents of the grid code or other grid regulation may differ between different Member States, but typically include the following provisions that are key drivers for the reliability of the electrical grid:

— Operation and performance characteristics of the transmission system;
— Technical requirements for the new and existing generators connected to the grid;

---

[8] Voltage ranges that are normally acceptable on the grid may be wider than what the nuclear power plant SSCs can tolerate, so the grid system operator needs to have a clear understanding of the voltage ranges that are acceptable for the nuclear power plant.

[9] Particularly in deregulated electricity regimes, there may be legal obligations to treat all power generating companies equally and place similar, if not the same, technical requirements on all power generating units when possible.

— Procedures related to grid planning and grid operation, such as the requirements for communications and the exchange of data;
— Obligations of the stakeholders, such as the grid system operator, transmission system owner, distribution network operators and generating unit operators.

For all significant (large) generating units, the grid code usually includes technical, design and performance requirements covering the contribution of the generating units to grid reliability and security (e.g. frequency control, voltage control, reactive power capability, stability, islanded operation, grid restoration, electrical protections at the grid–nuclear power plant interface).

Some nuclear power plants may not have all the capabilities specified in the grid code, based on their design (e.g. for frequency control[10] or for fault ride-through). Where a nuclear power unit technology is identified as being not fully compliant with the grid code requirements, it is necessary to have a formal agreement from the electricity regulator and the grid system operator to acknowledge this non-compliance. Any such agreements need to be approved by the nuclear regulatory body.

3.3.2. Requirements established by nuclear operators

The reliability requirements for a nuclear power plant are included in the documentation submitted to the nuclear regulator.

Complementarily, in some Member States, several nuclear operators have teamed up or collaborated with leading power institutes to do the following:

— Develop position papers and common requirements, representing up to date nuclear operator needs;
— Support and direct new reactor design assessments;
— Encourage the nuclear utility community to engage in strong and efficient interactions with other stakeholders (e.g. regulators, vendors).

These common requirements, known as user requirements (see Refs [27, 28]), encompass comprehensive performance characteristics for nuclear power generating units. They provide generic guidance on design requirements, including the electrical system design basis, which emphasizes a harmonized approach to grid performance and reliability, in line with the expectations outlined in the grid or network codes.

The performance characteristics are mainly based on the state of the art of nuclear engineering and the latest international standards. Most of the characteristics are compatible with the grid code requirements for power generating units.

3.4. GRID SYSTEM PLANNING AND OPERATION CRITERIA

The most important provisions for grid reliability are relevant redundancy, diversity and independence, which have to be considered in the planning stages. These provisions need to consider both normal and anticipated abnormal conditions, as well as the performance of the whole system and its parts, including system performance against hazards and potential single, common cause or dependent/independent multiple failures. Therefore, it is essential and fundamental to establish planning and operational criteria in advance and to plan for anticipated or predicted operation schemes with adequate robustness.

---

[10] This is discussed further in Ref. [26].

Generally, the grid system operator uses specific technical and economic criteria for grid planning and operation. The aim of these criteria is to operate the grid in an adequate and secure manner. A typical set of such criteria will normally include the following:

— Assessment of available generation versus load for different, typical points of time (system adequacy);
— Calculation of the available capacity of interconnections with respect to the generator's installed power, system peak demand or steady state stability;
— Measures taken in operation to withstand disturbances (e.g. electric short circuits, unanticipated loss of system facilities, rapid changes in renewable energy sources) and the applicable criteria to recover (i.e. operational security);
— Analysis of system stability (i.e. behaviour of electrical parameters following different small and large disturbances);
— Analysis of possible grid operational schemes and their robustness against different scenarios, including various uncertainties;
— Analysis of the ability to limit the degradations and to recover after unanticipated events or extreme events (i.e. grid resilience);
— Protection systems philosophy and criteria;
— Criteria for the availability of telecommunication networks.

For a specific electrical grid, these operational and planning criteria are usually specified in detail within the grid code or other similar requirement documents, discussed in Section 3.3.1.

From a nuclear power plant perspective, all these criteria are important to guarantee that the grid will support the following:

(a) Electricity export from the nuclear power plant through the electrical grid in all possible grid operating conditions;
(b) A high level of power quality for normal operation of all nuclear safety systems during reactor startup, normal operation and shutdown;
(c) Nuclear power plant house load supply when the nuclear power unit is not generating.

One important criterion is having the necessary number of redundant connections to ensure that individual unplanned events that cause the loss of transmission circuits do not lead to a LOOP. This is generally known as an $N - k$ criterion, described in the Glossary and in Section 4.1.

3.5. OPERATIONAL AND RELIABILITY DATA

A large volume of data is exchanged between the grid system and nuclear power plant operators during project development, construction, commissioning and operation. This section focuses on data that are important for reliability at the design and operation stages.

3.5.1. Data collected from nuclear power plants

The grid system operator needs to collect data regularly to be aware of the characteristics and performance in operation of the nuclear power plant. These data are mainly used by the grid system operator for grid operation, planning and development.

A set of data associated with the reliability of a grid for, and with, a nuclear power plant is given in Ref. [6]. Overall, the reliability data that are particularly related to the nuclear power plant and grid connection and operation can be placed into four major categories:

(a) Electrical parameter responses during normal operation and during different types of dynamic event (e.g. steady state frequency deviation, voltage ranges at connection points, normal and random load variations, power flow via the tie lines);
(b) Unavailability of important transmission system components (e.g. transformers, circuit breakers in power substations) and other elements that necessitate scrutiny;
(c) Availability, quality and behaviour of protection and communication systems;
(d) Power and energy reserves within a system and load shedding/outages impacting different types of transmission users (e.g. generating units, consumers, traders).

Grid system operators also need to be informed about the performance capability (designed and observed) of the generating units to perform the assessment and modelling of grid behaviour and reliability. Typical generator performance parameters collected and analysed include the following:

— The number and duration of planned and unplanned outages;
— The range of system voltage and frequency for which the generator can operate normally at full power;
— The wider range of voltage and frequency for which the generating unit can operate for a limited time without tripping, perhaps at reduced power;
— The worst transient events (e.g. short circuit fault close to the generator) for which the generating unit can continue to operate without tripping and without losing synchronism;
— The time needed for the generating unit to start up from a shutdown condition;
— The rate at which the generating unit can increase or decrease output on instruction;
— The ability of the generating unit to operate in automatic frequency responsive mode and the magnitude and rate of change of output that can be achieved;
— The active/reactive power diagram of the generator.

3.5.1.2. Data concerning nuclear power plant operation

The grid system operator expects from the nuclear power plant operator a realistic declaration of the status and values of operational parameters in the form of expected generation plans, values for rated or minimum power, or temporary limitations. The data declarations are often updated daily (as, for example, the net rated power depends on the temperature of the cooling source).

The grid system operator monitors and remotely collects values of electrical parameters at the common points of coupling and at the generator terminals. The status of some relevant equipment (e.g. breakers, control modes for active and reactive power) is also monitored.

3.5.2. Other reliability data collected

The collection of reliability data is essential for assessing the performance and robustness of the electrical grid and its interface with nuclear power plants. These datasets provide insights into system behaviour under normal and disturbed conditions, enabling operators to evaluate generation adequacy, frequency and voltage stability, and the effectiveness of protection schemes. In addition, monitoring disturbances at the grid–nuclear power plant connection point and tracking the operational status of grid components help to identify vulnerabilities and improve coordination between plant and grid operators.

Such comprehensive data collection supports informed decision making and enhances overall system reliability and resilience.

3.5.2.1. *Data concerning system frequency*

Quasi-steady state frequency data are collected to evaluate generation adequacy versus electricity demand within the electrical system.

Dynamic frequency response following different events such as short circuits, generator outages, load outages and the tripping of tie lines is used to evaluate the adequacy and efficiency of the load and frequency control reserves and the load shedding schemes, together with the underfrequency and overfrequency protection systems.

3.5.2.2. *Data concerning voltage*

Voltage data at each network node are collected and used to estimate the adequacy of the reactive power production capabilities.

Dynamic voltage data are used to estimate the efficiency of the reactive power reserves within certain areas, to verify the response time of protection systems in fault conditions, and to evaluate the ability of generators to stay connected and to maintain synchronous operation during low voltage conditions at the connection point.

3.5.2.3. *Data concerning disturbances at the grid–nuclear power plant interface*

It is important to install and operate an electrical fault recorder at the point of common coupling between the nuclear power plant and the electrical grid in order to be able to share transient analysis with the grid system operator.

Typically, fault recorders sample at high frequency and record voltages, active power, reactive power, power factor, frequency, harmonics, status of substation equipment (e.g. breakers, protections) and relevant signals, protections and parameters elaborated by the nuclear power plant.

While establishing the initiating events and chronology for such disturbances, it is also necessary to understand how and why the disturbance grew from one or two localized faults to something that had major consequences or that affected a wide area. A thorough analysis of such events can indicate that it is necessary to improve the arrangements for management and control of the system.

3.5.2.4. *Data concerning grid components*

Power flows need to be monitored because each network element has its rating or maximum allowable value of current and voltage to which it may be exposed. For lines and cables (including overhead, underground and submarine), the nominal current is of special interest, while for power transformers the apparent power is observed.

The status of the alarm and protection systems of the grid elements is also particularly important, considering fault clearances and grid robustness.

The connection between the nuclear power plant and the grid needs to be designed with an appropriate number of electrical circuits with sufficient ratings. The number of circuits needs to be determined according to all possible operating conditions, including the consideration of potential, possible or anticipated outages (usually of one or even two circuits simultaneously). The loadings of distant lines and transformers also need to be observed, especially if they are important for the stable operation of the nuclear power plant.

3.5.3. Grid fault records

A wide of range of hazards can lead to faults on a transmission system (see Section 43). The types of hazard that most frequently lead to faults will vary between Member States, depending on the geography, climate and the system design and maintenance practices. In order to optimize any plans for improving the performance of the grid system, the following data need to be collected and analysed for all faults on the transmission system, whether or not the fault caused disconnections of electricity consumers or generating units. The information that is normally collected includes the following:

— The circuit that experienced the fault and its location;
— The weather conditions at the time;
— The likely cause of the fault;
— The time it took to restore the faulted equipment to service;
— Whether any components suffered any significant damage;
— Whether the fault caused a loss of supply to consumers or the disconnection of the power plant;
— Whether the fault caused a cascade tripping of more circuits than planned in the design.

4. DESIGN AND OPERATIONAL ELEMENTS OF A RESILIENT ELECTRICAL GRID FOR AND WITH NUCLEAR POWER PLANTS

The electrical system, by its nature, is constantly subject to different statuses, environments, drivers and risks that can be grouped into four main actualities:

(a) Because of the non-storable nature of electricity, it is necessary to ensure the balance of supply and demand at any time. The aggregated demand of each electricity consumer is the image of the economic and social behaviour of the country, and this demand drives the control of the system. This image has a generally predictable dynamic shape, but with noticeable random deviations. It integrates many individual behaviours. It is influenced, even in the short term, by multiple factors, particularly the climate and the social habits of consumers.

(b) The electrical system is geographically widespread and has a strong relationship with the environment and infrastructure. Hence, it could face local or widespread weather events (e.g. lightning, storm, frost, flood, drought), which are quite predictable; however, their possible consequences have to be considered.

(c) System components are statistically susceptible to random or low probability failures, as well as unpredicted or catastrophic equipment failures, such as those caused by external events (e.g. impact from a vehicle, malicious acts by people) or ageing by known or unknown mechanisms.

(d) As with any complex systems managed by humans, undesired events triggered by human activity (or error) could occur during design, operation and maintenance. These errors may have acute impacts on the network; however, because of the close interdependence of the parts of the electrical grid, the impact may propagate with significant effects.

These actualities require the building up of security and operational margins as well as engineered and administrative controls in grid design and operation to prevent and protect against anticipated events and major failures, as well as to mitigate (and recover from) such major events and failures.

Therefore, to minimize the impact on the reliability and resilience of the electrical grid, it is important to do the following:

— Design and operate the system with some redundancy so that the more common unplanned events (faults) do not have significant effects.
— Identify the hazards or contingencies affecting the grid, their likelihood, their credibility, their possible impact, and whether they can be anticipated or not.
— Understand the possible failure modes propagating the effects of these hazards.
— Develop a strategy based on the defence in depth principles to minimize the impact on reliability and to limit the propagation of the degradations.

Section 4.1 describes the principles of establishing redundancy to minimize the effect of individual faults. Section 4.2 describes the range of hazards that can lead to faults and suggests ways to minimize their likelihood or effect. Section 4.3 describes ways of managing major faults that can occur following multiple individual faults, and Section 4.4 describes the principles of defence in depth. Finally, Section 4.5 describes recovery from major blackouts.

4.1. GRID SYSTEM DESIGN AND OPERATIONAL PLANNING CRITERIA

To ensure reliable operation in most circumstances, grid systems need to be designed and operated with some redundancy, so that individual unplanned events (faults) that cause the loss of a transmission circuit or a generating unit do not have a significant effect. This is generally known as the $N{-}k$ criterion, the simplest form of which is known as the $N{-}1$ criterion. For $N{-}1$, N refers to all the circuits in service and -1 refers to one item (e.g. a single circuit or component or a single generating unit) being removed from service as the result of an unplanned event. This is analogous to the single failure criterion used in safety systems in nuclear power plants. There may be an obligation placed on the grid system operator by the electricity regulator to comply with such a standard.

To use this criterion, it is necessary to define the types of unplanned event that need to be considered, and the types of consequence of these events that are unacceptable. The choice of unplanned events and their consequences will be different in different Member States, depending on the relative likelihood of the unplanned events and the cost and difficulty of ensuring compliance with the criteria. For example, for the $N{-}1$ criterion the types of unplanned event to be considered could be the following:

— The loss of one overhead line circuit;
— The loss of one underground cable circuit;
— The loss of one large transformer;
— The trip or disconnection of any one generating unit;
— A fault on one section of the busbar within a substation.

The types of unacceptable consequence could be as follows:

— Voltage at any point on the network outside the permitted range;
— System frequency outside the permitted range;
— Unacceptable overloading of any circuit or transformer;
— Disconnection of more than a specified amount of generation;
— Disconnection of more than a specified amount of demand;
— Loss of synchronism (pole slipping) of any generator;
— Cascade tripping.

These criteria lead directly to several design features. For example, the requirement not to disconnect more than a specified quantity of generation or demand on a single fault implies that large power stations (such as nuclear power stations) and large concentrations of demand need to be connected by multiple circuits, and that two or more large generating units are not connected to the same section of the busbar in a substation. The requirement to keep the frequency within a required range implies that there is sufficient frequency containment reserve in service to protect against the loss of at least the largest single generating unit, or the largest loss of generation from a single fault.

Some grid system operators use expanded security criteria of more than one element, defining that the consequences described above are not reached even in a case of simultaneous failure or outage of two elements ($N-2$ criterion) or the outage or failure of one element while another nearby is out of service for planned maintenance ($N-1-1$ criterion). For example, the unplanned events to be considered could include the following:

— Simultaneous loss of both circuits on a double circuit overhead line;
— Fault on a circuit breaker in a substation (which can cause the loss of two sections of the busbar);
— Simultaneous trip or loss of two generating units.

In order to fulfil the $N-1$ criterion continuously during the operational time frame, the grid system operator needs to take action after a fault occurs, to try to re-establish safe grid operation as soon as possible (i.e. to prevent a subsequent fault from causing one of the unacceptable consequences). From the nuclear power generating unit's perspective, it is important that the grid system operator informs the nuclear power plant operators promptly if such degraded security conditions persist for an extended time so that the nuclear power plant operators can take any actions necessary to maintain the safety of the nuclear power plant.

The $N-1$ criterion is generally used for the design of the connections for new power plants, and in some Member States the $N-2$ criterion has been used for the design of transmission lines connecting nuclear power generating units to the grid system.

Some Member States have a transmission system that is still developing and is not yet able to operate to the $N-1$ criterion for all typical faults. If such countries wish to install a nuclear power plant, then it is important that the grid performance be improved before the nuclear power plant is connected and commissioned to be able to satisfy the single failure criterion in future operating conditions. If this criterion cannot be met, then transmission system faults even far from the nuclear power plant could lead to cascade tripping and local system blackouts, thus increasing the probability of LOOP events.

The $N-1$, $N-2$ and $N-1-1$ criteria are usually applied deterministically (i.e. they are applied without detailed consideration of the probability of particular faults). However, expanded planning and operational criteria can be used that do take account of the probability of particular failures. One example would be to end certain transmission circuit outages and return circuits to service when severe weather is expected (since multiple faults are more likely during a period of severe weather, as discussed in Section 4.2). Some grid system operators also use probabilistic grid planning criteria, designing their grid to take account of certain events with low probability but with very significant consequences — for example, an event that disconnects a large amount of consumer demand or disconnects multiple generating units simultaneously.

4.2. HAZARDS AFFECTING THE RELIABILITY AND RESILIENCE OF THE ELECTRICAL GRID

There are many possible kinds of hazard that can challenge the reliability of a transmission system, leading to electrical grid faults and disturbances[11], some natural and some human-made, such as the following:

— Weather hazards (e.g. lightning, high winds, heavy rain);
— External hazards (e.g. floods, earthquakes, geomagnetic storms);
— Hazards from flora and fauna;
— Equipment failure;
— Human hazards, including human error and accidental or malicious damage.

The relative significance and frequency of occurrence of these various kinds of hazard can be very different in different geographical and demographical areas. Hence, the number and significance of faults due to each cause — and the measures for prevention, protection and mitigation — can be very different between different power systems. Some hazards are anticipated by the electrical grid designers and operators, and measures are taken against them. Some other hazards are of high impact and low probability, such as extreme events or a combination of multiple anticipated events, where the mitigation measures are not practicable because of the high cost. The following sections discuss common hazards and measures against such hazards. (The frequency and severity of events related to weather may change in the future because of the climate change. Such changes are discussed in more detail in Section 8.3.)

4.2.1. Natural hazards for the transmission system

The reliability of the transmission system can be significantly challenged by a wide range of natural hazards. Events such as high winds, lightning, heavy precipitation and floods, cold weather, fog, biological intrusions, extreme heat, fires, earthquakes and geomagnetic storms can damage transmission lines, towers, substations and associated equipment, or degrade their performance. These hazards may cause faults, reduce equipment margins, hinder access for repairs or trigger cascading failures, ultimately affecting the secure power supply to nuclear power plants. Understanding these threats and implementing appropriate design, maintenance and operational measures is essential for ensuring system resilience.

4.2.1.1. Wind

The experience in Europe and North America is that many major loss of supply events are due to high winds, severe storms, hurricanes or tornados (see Refs [34–39]). In extremely severe storms, there is significant damage to the transmission networks, with the extent of damage increasing as the square of the maximum wind speed.

High winds (particularly when combined with other hazards) can lead to faults and damage to the elements of the electrical grid directly or indirectly, such as the following:

— Overhead lines swing or oscillate in high winds (these large vertical oscillations are commonly termed 'galloping') so that live conductors come close enough to allow a flashover or touch. This can be worse in freezing conditions if a layer of ice builds up on the conductors (as described in Section 4.2.1.4).

---

[11] There are various definitions and classifications of faults and disturbances in different Member States (see Refs [30–33]). For the purpose of this publication, any unplanned event that causes a circuit or an item of equipment in a network to be switched out of service, either automatically by the electrical protection system or manually in response to an alarm signal, is considered a fault.

— Flashover faults without damage to overhead lines can also be caused by windblown debris or swaying of trees beside the overhead lines.
— In higher winds, overhead lines and insulators can become damaged by larger windblown debris or by trees falling onto the overhead lines.
— In extremely high winds, transmission towers themselves can become damaged and buckle.
— For power systems near a sea coast, an additional hazard due to prolonged high winds is the gradual buildup of salt on the insulators of overhead lines or substations near the sea. If this is followed by damp or humid weather (and fog) (as described in Section 4.2.1.5), there can be frequent flashover faults. Furthermore, salt deposits on exposed cables and metal parts may lead to corrosion and possible short circuits.
— For power systems with a sea coast, storm surges due to extreme winds may cause coastal flooding of substations, transmission tower foundations and support structures (see Section 4.2.1.3).

It may be possible to improve the reliability of transmission lines and towers that are important for the connection to a nuclear power plant by:

— Reducing the risk of direct wind damage by planning their routes to avoid areas with higher wind speeds, such as across mountain ranges;
— Designing grid structures to withstand the maximum projected wind loadings, including the possible combination of natural hazards (such as added loading of ice and snow), meeting or exceeding the requirements in published standards (e.g. Ref. [40]);
— Vigorous management of trees growing near overhead lines (see Section 4.2.1.6).

4.2.1.2. Lightning

A lightning strike causes destruction by direct impact and propagates overvoltages or overcurrents causing damage, short circuits and voltage surges. Furthermore, lightning may create false or spurious signals in instrumentation and controls at the plant or on the grid (see Ref. [6]).

The goals of the lightning protection systems and arrangements are as follows:

— To control the location of the probable impact points and to guide the currents (e.g. using surge arresters, cages, cables);
— To divide and to diffuse the currents in the ground (i.e. grounding systems);
— To limit coupling and propagation of overvoltages and overcurrents (e.g. using shielded cables, grounding connections, spark gaps).

The lightning protection of the substation connected to the nuclear power plant, as well as the associated high voltage towers and circuits, needs to be carefully evaluated to determine the remaining voltage margins following a lightning strike on the transmission lines. The choice and installation of lightning protection systems have to be based on the latest international standards. Some ways of improving the protection of overhead lines and associated equipment from lightning strikes are described in Refs [41–44].

4.2.1.3. Heavy precipitation, floods and landslides

Very heavy rain occasionally causes flashover faults (short circuits) across insulators, but in regions where this is a problem, using an alternative design of insulator can reduce this risk (see Ref. [45] for further details). Heavy rain can also cause water ingress into high voltage insulators and switchgear, leading to internal flashovers and catastrophic failures. Careful maintenance can minimize the risk of such short circuits.

More significant consequences of a prolonged period of heavy precipitation are flooding, landslides and avalanches. Floods can also be caused by high winds or storm surges that damage flood defences. The main threat from flooding is to equipment such as protections, switchgear, transformers and control cubicles mounted at ground level in substations. Flooding does not generally pose a threat to overhead lines, although there is the potential for weakening transmission tower foundations and support structures.

Landslides or avalanches in mountainous areas can occur following very heavy precipitation and can damage underground cables or transmission towers supporting overhead lines. However, the greatest impact would result from damage to a major substation.

The main remedial action against flooding or landslides for new installations is to avoid locating them in areas that are at risk. For existing installations in at risk areas, it is good practice to reconsider the relevance of the criteria associated with these risks and to carry out routine reassessments of the consequences to establish the necessary defences.

4.2.1.4. *Cold weather*

Heavy snow and the accumulation of ice can cause failures of overhead lines or towers because of their weight and size, particularly when accompanied by wind. Some examples of possible cold weather effects are the following:

— A layer of ice increases the cross-sectional area of the overhead line conductors, thus increasing the sideways forces, and hence the risk of collapse, in high winds.
— The weight of snow can cause branches or whole trees to fall and damage lines.
— Under certain wind conditions, ice can build up on overhead line conductors to form a streamlined aerofoil shape, which can lead to large vertical oscillations (i.e. galloping), causing fatigue failures of the lines or supports (see Refs [41, 46]).
— An extreme ice and snow event is an ice storm, which occurs when supercooled rain, in combination with a strong wind, freezes on contact with trees or structures, rapidly forming a thick layer of ice. Some examples of such events were observed in Canada and northeastern USA in January 1998 [47–49], in Germany in November 2005 [50] and in Slovenia in February 2014 [51]. In all these events, many transmission towers were damaged.

In regions likely to experience ice and snow, it is normal practice to design the transmission towers and lines to allow for the maximum expected ice and wind loading, but there is a limit to what is practicable. For example, in the 1998 event in Canada, ice reached a thickness of 70–90 mm on overhead lines, which was much greater than the thickness allowed in the design (see Ref. [47]) or in national or international standards.

Large sized hail stones during severe hailstorms can also damage outdoor and exposed elements of substations or structures that are housing equipment or insulators on transformers and overhead lines. Indirectly, they can contribute to flooding by clogging drains.

Flooding, landslide, excessive snow or avalanches would also inhibit transportation and movement of critical personnel and materials to where system equipment is damaged or submerged. It is likely to take many weeks to repair or replace equipment, increasing the risk of multiple secondary faults.

4.2.1.5. *Fog and dew*

The ceramic or glass insulators on overhead lines and in substations can experience surface contamination from a variety of causes, including the following:

— Windborne salt from the sea;
— Windborne dust or chemical pollutants;
— Dried bird excrement.

These forms of pollution are generally not good electrical conductors when dry but can become good conductors when damp. This can happen in foggy conditions or following a heavy dew in the early morning. When this happens, there can be frequent flashover faults across the insulators affected, which can continue until heavy rain cleans the insulators. Such foggy conditions have been the initiating cause of several major blackouts, for example, in northern India and Pakistan (see Ref. [52]).

The possible remedial measures for such problems include using insulators with longer tracking distances or different profiles (see Ref. [41]), applying a grease or polymer coating to ceramic or glass insulators or using composite polymeric insulators. For substations, it may be possible to provide a means of washing insulators. Leakage current sensors can identify when washing is required, allowing condition based maintenance.

4.2.1.6. Flora and fauna

Overhead lines and substations need to be physically protected against the effects of flora and fauna.[12]

It is important to manage trees and other vegetation growing near or underneath overhead lines to avoid flashover to the vegetation or trees. Governments in Member States need to ensure that the transmission operators have sufficient legal powers for the necessary tree cutting. Hazards related to trees may also be reduced at the design stage by:

— Rerouting overhead lines to avoid forested areas;
— Replacing overhead lines in forested areas with underground cables;
— Using a tower design with no cascading fall mode (with which the fall of one tower will have a very low probability of causing the fall of neighbouring towers).

In some areas, it may be necessary to secure the perimeter of substations against the intrusion of animals and to take pest control measures to eliminate the possibility of animals chewing on cables. It is beneficial to ensure that all electrical enclosures are proofed against rodents or other small animals.

In some Member States, large birds (e.g. storks, cranes, vultures) are a major cause of faults on the transmission system when they perch on transmission towers. Experience with such faults and possible remedial measures are described in Ref. [53] (e.g. placing antibird spikes on cross-arms (horizontal members) at the top of transmission towers to prevent birds perching there).

4.2.1.7. Hot weather and droughts

High temperatures raise the operational temperature of grid components, including their physical environment.

A long period of dry weather can cause the ground to dry out, which reduces the thermal resistivity of the soil, reducing the margins of the rating of underground cables. Drying of the subsoil can also reduce its electrical conductivity, which needs to be considered when designing earthing systems (see Refs [54, 55]).

A more significant effect of a drought is that vegetation dries out and increases the risk of fires. Overhead lines can initiate wildfires in drought conditions if vegetation comes close to, or in contact with, the overhead line conductors, causing flashovers.

Higher temperatures of grid components than those considered at the design stage induce short circuits, degrade design margins, accelerate ageing and cause higher transmission losses by, for example:

— Increasing the sag on overhead lines owing to thermal expansion and hence reducing the clearance distance to items below the lines, thus reducing the maximum power that the lines can safely carry;
— Increasing the risk of flashovers to structures below the line or to trees, as described in Section 4.2.1.6;

---

[12] The fauna also need to be protected against electrical shocks.

— Reducing the cooling capacity of equipment such as transformers, thus limiting the maximum power that they can carry or causing them to exceed the temperature limits set at the design stage;
— Increasing energy loss in the system because of higher electrical resistance, although the effect is relatively small.[13]

The upper conductor temperature limit of overhead lines can be reduced by increasing tension to reduce the sag or by using conductors with a lower coefficient of expansion. The temperature of transformers and high voltage cables needs to be monitored routinely and particularly during hot temperature episodes. The upper ambient temperature limit for transformers can be increased by installing larger or additional coolers. Since high voltage cables can have an expected life of more than 40 years, it is important at the design stage to consider raising rating temperature factors and margins in layouts.

4.2.1.8. Fires

Smoke from fires can cause repeated arcing faults on an overhead line because the ionized air in the smoke can become a conductor of electricity. When a fire occurs near an overhead line, the conductors and insulators can become damaged directly by heat. To reduce the risk of damage from fire in areas that experience prolonged droughts and hot weather, it is necessary to control the vegetation under and near overhead lines (see Section 4.2.1.6).

Forest and wild brush fires can also disrupt the ability of critical personnel and emergency responders to access the plant and equipment. Hence, it is good practice towards improving resilience to develop plans for emergency access to allow the timely repair and restoration of the electrical grid in case of nearby wildfires.

4.2.1.9. Earthquakes

In areas where strong earthquakes occur, the general experience is that the design of transmission towers for resistance to the horizontal forces due to high winds appears to be adequate to prevent direct damage to towers by earthquakes (see Refs [57, 58]). However, transmission towers can sometimes become damaged or be toppled because of an earthquake induced landslide.

Earthquakes can cause significant damage in substations. Earthquake induced damage commonly includes physical damage of the porcelain members of high voltage equipment and broken or leaking bushings. A lack of adequate slack in conductors connecting equipment can also load and damage bushings and post insulators because of motion induced by the earthquake.

A significant aspect of damage to substation equipment is the limited amount of spare or replacement equipment that may be immediately available and the long time required to replace or repair damaged high voltage equipment, even if spares are available (see Refs [57, 58]). Some guidelines are available to assist in the design of substation equipment for more resilience to earthquakes in Refs [57–59].

4.2.1.10. Geomagnetic storms

Geomagnetic storms, in general, are most intense around a peak in the sunspot cycle (approximately once every 11 years). Major geomagnetic events occur once in every 100 years, but with little or no warning.

Such storms can cause large currents to flow in the upper atmosphere near the Earth's poles, inducing currents (i.e. geomagnetically induced currents) on the surface of the Earth, particularly near the polar regions. This phenomenon is described in Refs [60, 61]. They also induce unusual currents in suitable conducting structures, such as overhead lines and circuits in electrical grids. Although the effects of such storms are likely to be greater nearer the poles, the effects may spread to larger areas via the

---

[13] A 2.5°C increase in the temperature of aluminium and copper conductors results in a 1% increase in the resistance, and hence a 1% increase in the power loss (see Ref. [56]).

electrical grids. Long distance overhead lines of grids span large geographical surfaces along the direction of the currents in the upper atmosphere, acting as conductors for the geomagnetically induced current (see Ref. [62][14]).

The geomagnetically induced currents are of very low frequency, sometimes described as quasi-direct current (quasi-DC), and would be similar in the three phases of a three phase transmission system (i.e. a zero sequence current). The potential effects on electrical systems include the saturation of the cores of transformers and the overheating of ground connection conductors or resistors. The magnetic cores of power transformers of five limb construction, or those with separate single phase units, may saturate. (This would be unlikely to happen for three phase power transformers of three limb construction.) Saturation of the cores of large transformers may cause extensive damage to these cores, as described in Ref. [62], and may cause large harmonic currents, which may result in overheating damage to generators and large motors. The damage to transformers may result in long term outages of large portions of the electrical grid. Saturation of the cores of protection transformers can also lead to spurious protection operation, which could cause nuclear power plant generator trips and reactor shutdowns. During such an event, power plant operators and the grid system operator are likely to see large swings in indications of system voltage and reactive power flow.

Where the risk of geomagnetic storms is considered high (e.g. near the polar regions), the mitigation measures may include passive devices such as neutral current blocking devices (capacitors) in the common neutral to ground link of the three phase transformers, or circuit modifications as described in Ref. [63]. Mitigation can also be provided through operational procedures, such as taking certain assets out of service if they are at risk of being damaged, or implementing rapid islanding or reacting aggressively to avoid voltage collapse (see Refs [60, 61, 63, 64]). Training also needs to be provided to improve awareness of geomagnetic storm risks, the likely consequences and the appropriate operator response.

4.2.2. Equipment failure

If the transmission system is designed and operated to have some redundancy and diversity such as complying with the $N-1$ criterion described in the Glossary, then the failure of a single item of transmission equipment (e.g. a transformer, a circuit breaker) would not normally have a significant effect on the overall reliability of the transmission system.

However, it is necessary to take adequate measures to avoid or limit the propagation of the consequences of a single failure. For instance, equipment in the vicinity of transformers needs to be protected against fire and projectiles from failure of the transformer, and circuit breakers have to be protected from catastrophic failures of nearby circuit breakers. Overhead line distance protection has to be duplicated or have backup protection using a different type of fault detection.

The consequences of equipment failure will be greater if the failure occurs at a time coincident with another fault on the system. Examples could be if a circuit breaker fails to open to clear a fault on an overhead line, or there is a malfunction on a protection system so that a fault is not cleared, or more equipment is switched out of service than is necessary following a fault. An example of protection malfunction is described in section 13.5 of Ref. [6].

---

[14] Regarding the effects of an unexpected geomagnetic storm in North America that caused a widespread event starting with the Hydro-Québec power system, Ref. [62] notes that the "event on the Hydro-Québec power system…resulted in its complete collapse within 92 seconds, leaving six million electricity consumers without power. This same storm triggered hundreds of incidents across the United States including destroying a major transformer at an east coast nuclear generating station." It is also estimated in Ref. [62] that "Should a storm of this magnitude strike today, it could interrupt power to as many as 130 million people in the United States alone, requiring several years to recover."

4.2.3. Human error

During the initial installation or during repair and maintenance activities, pieces of equipment may be assembled incorrectly, and will therefore not operate as intended in a future event. Errors may also be made in the settings of electrical protection equipment, so that either it operates when it should not or it fails to operate when it should. A particular concern with electronic multifunction systems is to ensure that any features that are not wanted are correctly disabled. The main remedial measure against such a human error is an adequate quality assurance programme covering the non-nuclear safety qualified SSCs of the nuclear power plant and of the grid–plant interface, and all operations and maintenance activities.

Hazards can also arise from third party actions (i.e. actions by members of the public who are not directly involved in the operation of the electrical grid). For example, a tall vehicle or machinery, such as a crane, could be driven under or operated near an overhead line and come within the safety clearance distance, or make contact, causing a flashover from a live conductor to ground. The risk of such hazards to overhead lines may be reduced by ensuring extra clearance distances on overhead lines near public highways or by barricading an area where cranes are used. Where overhead lines cross agricultural land where machinery may be used, it is important to increase the awareness and knowledge of farmers and provide clear warning notices at the entrances to fields.

4.2.4. Deliberate or accidental destruction of system elements

Accidental or deliberate damage to transmission equipment can cause failure or may require circuits to be switched out of service for safety reasons.

There have been examples of accidental damage to transmission system components. For example, hunters shooting at birds have caused collateral damage to the glass or porcelain insulators on overhead line circuits or to the fibreoptic cable that is part of the ground wire on an overhead line circuit that forms part of the communications for electrical protection.

Regardless of the severity of deliberate attacks, the legal system in each Member State needs to provide for suitable punishment for individuals who are found guilty of causing malicious damage to essential parts of national infrastructure, such as the public electrical system and connections to a nuclear power plant, to provide a suitable deterrent.

4.2.4.1. Metal theft

Several Member States have experienced increasing rates of theft of metal, including theft of metal that is important to the safe and secure operation of the transmission system. Examples include the theft of the following:

— Parts of the copper ground material from a substation or the copper grounding straps from transmission towers;
— Parts or components of substation equipment when it is out of service for maintenance or repair;
— The ground wire from overhead lines;
— Steel parts from transmission towers.

In addition to the cost and time delay for the replacement of stolen parts, time may be needed for the repair of any equipment damaged during the theft. To deal with the theft problem, some Member States have introduced legislation to discourage metal theft, by regulating the trade of scrap metal to make the sale of stolen metal items more difficult (e.g. the UK Scrap Metal Dealers Act [65]).

An attack — physical, cyber or blended — on system elements may lead to the equipment failure(s) discussed in Section 4.2.2 as beyond the design criterion and hence to the complete collapse of large parts of the electrical grid.

Physical and computer security protections need to be provided against coordinated attacks on key elements of the electrical grid. In a graded approach, some substations may need reinforced security provisions selected on the basis of a security risk analysis. These may include substations that connect to nuclear power plants or those with multiple circuits or large power flows. To mitigate the consequences of malicious attacks, it may be necessary to retain a national or regional reserve of critical spare parts. As an example, in the case of transformers, transformer stations are often located in various urban and rural areas and can remain exposed to attack, causing disruptions in the local and national power grid. Maintenance of spare transformer units and critical parts may mitigate the mid-term and long term consequences of a malicious attack against electrical grid infrastructure.

The operation of an energy transmission system relies on wide digital instrumentation and control systems, some of them remote, providing a large surface at risk of cyberattacks. The grid system operator needs to have a continuous overview of the system's state, including information on the operating conditions of all power plants and major transmission circuits, and needs to communicate with the power plants to provide operational instructions. Many modern data, communication and control and protection schemes use digital technology — standalone or interfacing with network based systems. Hence, it is possible for an incorrect or corrupted data signal sent via the communication channels to cause incorrect operation of equipment. An incorrect data signal may be sent by accident or be due to an equipment fault, but may also result from a deliberate attempt to cause damage (i.e. a cyberattack).

For example, it may be possible to gain access to digital equipment in an electricity substation to execute malicious control either through communication networks or through local portals at substations intended for computer connectivity. To prevent such cyberattacks, sufficient attention has to be paid to the computer security of substations and grid control centres (see Refs [66–69]). The similar computer security arrangements for a nuclear power plant are also discussed in Ref. [69].

From the point of view of the nuclear power plant, utilities should apply the concepts in the IAEA Nuclear Security Series guidance, such as the establishment of a computer security programme (NSS 42-G) [70], computer security risk management and an appropriate defensive computer security architecture (NSS 17-T (Rev. 1)) [71] related to the interface between the nuclear power plant and the grid instrumentation and control.

Extensive and coordinated attacks on the electrical grid infrastructure, including the grid control centre, have to be assessed regarding the appropriate risk informed approach and threat analysis, and may result in important societal consequences. Those hazards are further discussed in Section 4.3.5.2.

4.3. MAJOR FAILURE PHENOMENA IN GRID SYSTEMS AND EFFECTS ON NUCLEAR POWER PLANTS

The prevention, protection and mitigation measures and controls introduced in Section 2 may not be sufficient if there are multiple, independent or concurrent events, such as the following:

— An individual weather event, such as a lightning strike, followed by a malfunction of a protection system;
— Multiple faults during a short period of time, such as a period of severe weather;
— Multiple faults with physical damage to the transmission system by human induced or natural events.

Regardless of their underlying causes, faults and failures may have consequences for the electrical grid that can be grouped into four particular event types according to their characteristics and impact:

(a)   Overload cascade;
(b)   Voltage collapse;
(c)   Abnormally high or low frequency;
(d)   Loss of synchronism.

These major grid phenomena may have implications for the stability of the connected generating units, including nuclear power plants. The severity of the impact can range from simple oscillations of power, voltage or frequency at the connection point to islanding or tripping of a generating unit. Each of these phenomena is described in the following subsections with the identification of major impacts on nuclear power plants and specific features and provisions, followed by further and more detailed discussion on the engineered and administrative measures and controls in Sections 6 and 7.

4.3.1. Overload cascade

An overload cascade is the phenomenon where an incident causes an element of the transmission system to be disconnected, which causes another element to be overloaded so that it is also disconnected. This disconnection causes more overloads of other elements, which in turn are disconnected, causing more and more overloads, in a chain reaction, with the following possible sequence:

— The initiating occurrence may be the consequence of several types of event, in particular, a sudden trip of a line or a generating unit.
— An unexpected change in demand may occur, which causes a change in power flow inconsistent with available overhead line capacity, possibly combined with low voltage.
— Eventual damage of overhead lines, underground cables or transformers that are operated for an extended period with excessive loading may be the final event of the cascade.

The general scheme of the overload cascade is shown in Fig. 2. As such occurrences are anticipated and their impacts are known, overload protections are normally installed and used to protect electrical conductors. For example, a small overload of a circuit will initially initiate a warning to the grid system

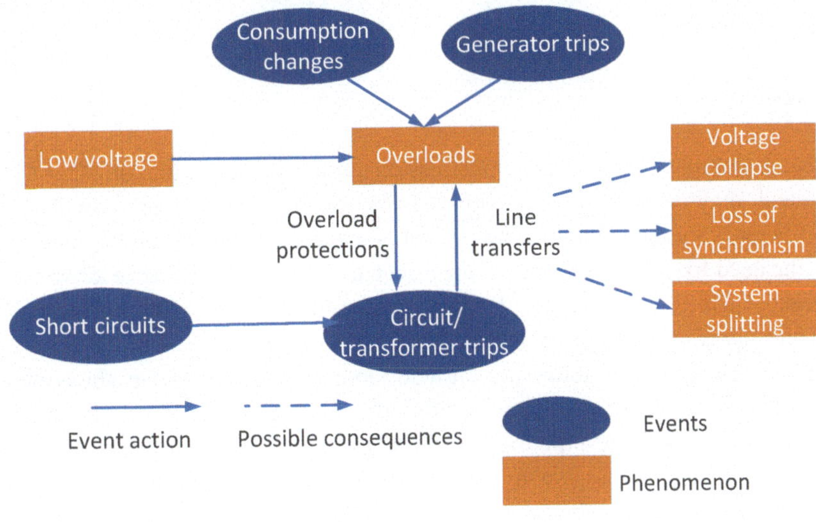

FIG. 2. Initiation and consequences of overload cascades.

operator, but if the overloading is not reduced within a certain period (e.g. 20 min), the overload protection operates automatically, tripping the affected circuit.

When a circuit is tripped by the overload protection scheme, the load that was previously transmitted by this circuit diverts to other circuits, increasing their loading. This may cause these other circuits to overload, triggering the actuation of their overloading protection schemes, causing these circuits to trip, and as a result overloading more circuits and causing more trips. As the overload condition cascades, control of the system becomes increasingly difficult, since the new power flows become increasingly difficult to redispatch and manage. This may eventually lead to increased power demand and decrease in voltage at the common points of coupling, as well as oscillation of the rotor angle of generating units with several potential consequences, including a voltage collapse (see Section 4.3.2) and/or a loss of synchronism (see Section 4.3.4) or a split of the system into several electrical islands owing to the actions of grid protection systems.

4.3.1.1. *Impact on nuclear power plant operation*

The impact of overload cascades on a nuclear power plant depends on the magnitude of voltage and power transients at the point of common coupling, as well as rotor angle deviations in the main generator. Step changes in voltage or power can lead to power oscillations, which in turn can cause erratic operation of plant equipment, such as slowing or speeding of pump motors, and then the activation of process or electrical protections in the plant electrical power system.

4.3.1.2. *Reliability provisions*

Variations in electrical parameters can induce the automatic activation of protections to disconnect the plant from the grid system on the basis of electrical parameter abnormality (e.g. on maximum/minimum voltage, pole slip, reverse power) or can automatically trip the plant on other reactor protection process parameters (e.g. reactor coolant pump overspeed, low departure from nucleate boiling, high linear heat rate, high neutron flux). The correct tuning of parameters such as turbine speed/power and generator voltage controls and limitations is very important to dampen oscillations to stabilize the turbogenerator.

The main reliability provisions against significant variations of electrical parameters (in terms of amplitude and frequency/duration of the deviations) are the adequate design and tuning of the turbine and generator controls and of the generator pole slip protections. The choice of such parameters needs to be carefully elaborated and documented because these parameters are mainly the result of a trade-off between underdamping oscillations and overprotecting the generator. Stability tests and conditions are often detailed in grid codes.

4.3.2. Voltage collapse

A failure that disconnects one or more transmission circuits or generating units, or a sudden increase in consumer demand, may cause an increase in reactive power demand in an area. This may cause the voltage to fall sufficiently for the voltage at the nuclear power plant's substation to go below the minimum level allowed. If the need for reactive power of the affected zone is beyond the reactive capabilities of the neighbouring areas — which are inherently limited — then the neighbouring areas may also experience low voltage, causing significant voltage drops on the transmission network. During voltage fall, operators may need to block the operation of automatic on-load tap changers; otherwise, they automatically raise the voltage and thus increase the demand, accelerating the voltage fall. Below a given voltage threshold, typically called 'critical low voltage', the limit of the transmittable power is reached, and if no action is taken, this leads to a voltage collapse (i.e. a fast drop of the local voltage, known as a 'brownout'). This phenomenon is shown in Fig. 3.

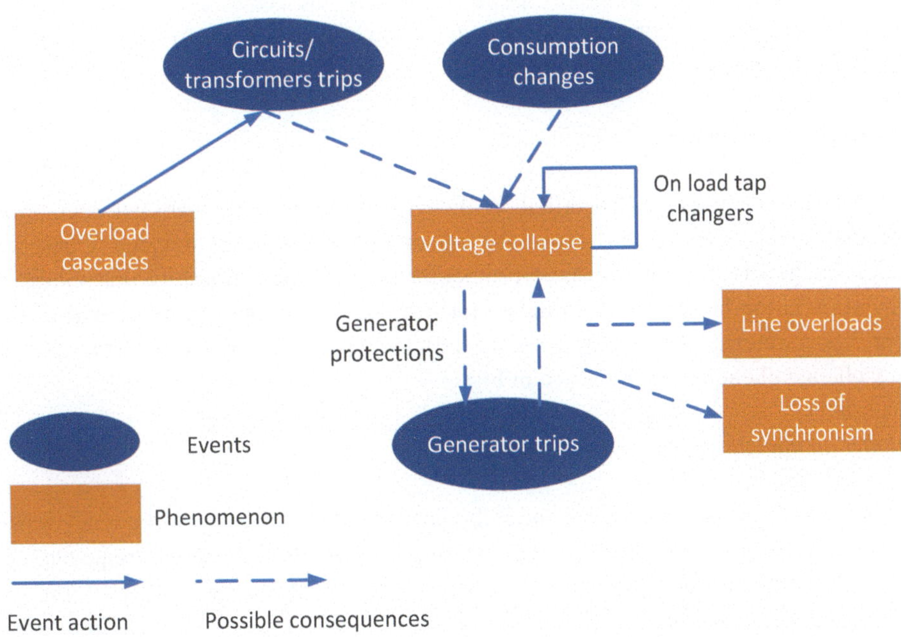

FIG. 3. Initiation and consequences of a voltage collapse phenomenon with likely sequences.

4.3.2.1. Impact on nuclear power plant operation

At the grid–nuclear power plant interface, when the grid system voltage starts to drop, the voltage controls (i.e. the excitation system) of the nuclear power plant generator will attempt to keep its voltage constant. The voltage regulator will drive the generator to deliver more and more reactive power until it reaches a limit. This limitation is typically implemented in the control systems that protect the generator against a reactive power overload.

If the grid voltage keeps fluctuating and dropping slowly (e.g. in a few minutes), then the undervoltage alarms alert the plant operator to closely monitor the voltage controls and the plant behaviour. If necessary, the plant operator will contact the grid system operator, who will have to take proper measures to stop voltage degradation. If the situation worsens, the plant operators may manually trip to house load operation at their discretion or upon a request from the grid system operator. If the unit does not have the capability to trip to house load, then the plant operators will typically manually trip the reactor and transfer to on-site power supplies.

If the voltage drops rapidly (e.g. in a few seconds) or below a low set point with a time delay, then the undervoltage alarms and, eventually, the protection actions are normally activated (typically 85% of the nominal voltage with a few seconds' time delay). This will cause the generating unit to automatically trip to house load or to trip the reactor and disconnect from the grid and transfer plant loads to on-site emergency power supplies (e.g. diesel generators), which will initiate and provide power to safety related equipment.

4.3.2.2. Reliability provisions

At the grid level, the transmission of reactive power over long distances is avoided, as that would cause increased transmission losses as well as large voltage differences between different points on the network and decreased transmission capacity used to transmit active power.

At the nuclear power plant level, the main reliability provisions are the assurance that the generator is able to deliver the full reactive power when needed by having adequate control and settings for the reactive power limiters. If there are fault conditions where the reactive capability of the generator would not be sufficient to maintain the grid voltage, it may be necessary to install additional reactive compensation equipment on the transmission system.

4.3.3. Rise or fall of frequency

Outside the frequency ranges specified in the grid code or other national regulations, serious unpredictable malfunctions can occur, especially on motors and control devices. Consequently, it is normal practice for generating units to have protection that will automatically disconnect them if the frequency deviation is too large and will almost certainly lead to collapse of all or part of the electrical system. Under these circumstances, the frequency can fall at a very rapid rate (i.e. a few hertz per second). The frequency change phenomenon is shown in Fig. 4.

4.3.3.1. Impact on nuclear power plant operation

If a frequency fall (or rise) is detected and if the turbine controls are set in frequency sensitive mode, then the power controls raise (or decrease) the power in proportion to the relative frequency deviation, up to a frequency deviation that is smaller than the maximum frequency steady state deviation.

If the frequency keeps dropping, the turbine control adjusts the power according to a predefined maximum frequency dependent power reduction curve. If the frequency reaches the lower limit of the frequency domain, the underfrequency protection relay (e.g. typically set at 47.5 Hz in a 50 Hz system, with no time delay) disconnects the plant from the grid, either tripping to house load or tripping the reactor and transferring plant loads to on-site emergency supplies.

If the frequency keeps rising and reaches the limit, the turbine control may reduce the power according to a predefined minimum frequency dependent power reduction curve, even if the plant is not in a frequency sensitive mode.

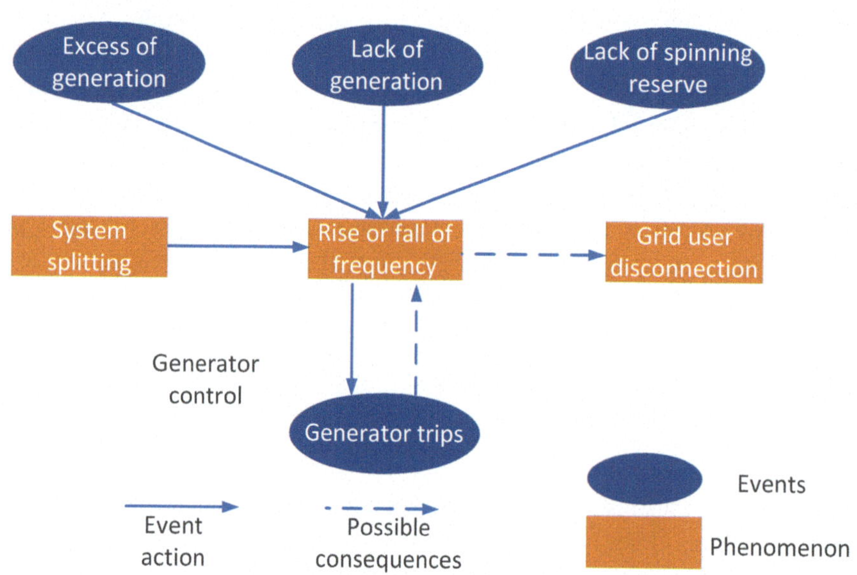

FIG. 4. Initiation and consequences of a frequency fall phenomenon.

4.3.3.2. Reliability provisions

For a nuclear power plant, the principle of protection against out of range frequency is to isolate the plant from the grid and reach the best safe condition when disconnected from the grid (house load operation or hot shutdown).

For the grid, when the frequency drops too far, there is usually an automatic load shedding for demand or a signal to increase the power sent to the generators. For high frequency, automatic signals may be sent to the generators to reduce power (mainly for gas fired and renewable power sources).

4.3.4. Loss of synchronism

The loss of synchronism of a nuclear power plant generator may be due to an external cause or an internal cause, such as the following examples:

— A short circuit fault on the transmission system nearby that is not cleared rapidly by the electrical protection;
— A trip of several transmission circuits or grid transformers that leaves the generator with a weak connection to the rest of the system;
— Failure of the excitation system in the generator.

The loss of synchronism phenomenon is shown in Fig. 5.

4.3.4.1. Impact on nuclear power plant operation

When a generating unit is pole slipping, there will be pulses of power every couple of seconds, with the generator alternately exporting and importing power, and the voltage on the grid near the generator varying from a high value to a low value. Pole slipping causes large varying currents to flow in the stator

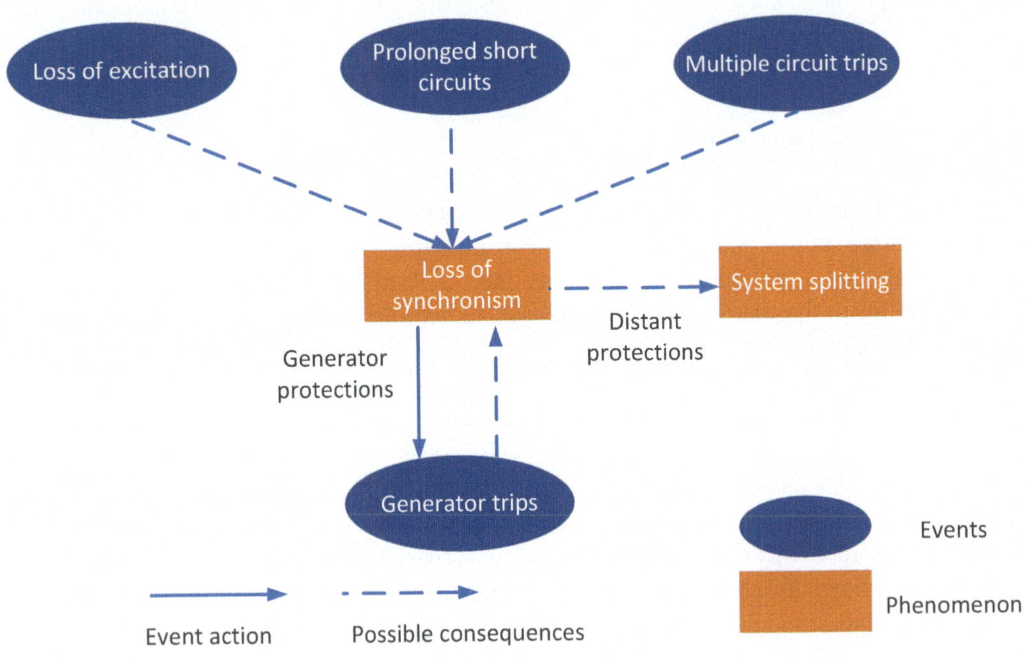

FIG. 5. Initiation and consequences of loss of synchronism.

winding and exerts large oscillatory forces on the generator rotor and stator and on the generator and turbine foundations. It can cause significant mechanical damage if it continues for long.

Nuclear power plant generators are normally equipped with the following:

— Loss of excitation protection, detecting the loss of the excitation function of the generator and tripping the reactor.
— Pole slipping protection, detecting and counting electrical angle turns between the nuclear power plant generator and the grid electrical field. After the count of a few turns, a loss of synchronism is detected and the protection trips the nuclear power plant to house load.

Variations of electrical parameters can induce other automatic activations of protection based on abnormality (e.g. on maximum/minimum voltage, pole slip, reverse power) or trip the plant on other reactor protection processes (e.g. reactor coolant pump overspeed, low departure from nucleate boiling, high linear heat rate, high neutron flux).

4.3.4.2. Reliability provisions

The main reliability provisions against the loss of synchronism phenomenon are adequate design and tuning of the turbine and generator controls as well as good coordination between the three protections against loss of excitation, pole slipping and reverse power. The choice of parameters needs to be carefully elaborated and documented because it is mainly the result of a trade-off between underdamping oscillations and overprotecting the generator.

The settings for this protection are the result of a trade-off between the need to protect the generator (supported by the nuclear power plant vendor) and the need to keep the plant connected to the grid as long as possible in the case of electrical transients (supported by grid system operators).

Protection thresholds and count durations need to allow discrimination of short transients experienced during fault clearances from synchronism perturbations coming from the grid. The nuclear power plant's protection needs to be coordinated with grid protections, which are designed to isolate the area where the origin of the loss of synchronism is detected.

For nuclear power plants with several units, it is necessary to set different sets of thresholds for each unit, first to maximize the probability for units that have not tripped to house load to reconnect quickly, and second to minimize the risk of having two or more simultaneous trips to house load.

4.3.5. Loss of grid system control or integrity

For the nuclear power plant operator, the most important aspect of the reliability of the grid control system for the nuclear power plant operation is the availability of the main functions of the grid control centre. Two issues that a grid control system may face are particularly important for a nuclear power plant: the availability of the control centre and the availability of communication channels (e.g. messages, data, voice) with dispatchers.

4.3.5.1. Loss of communication between the nuclear power plant and the grid control centre

A loss of communication may be partial (i.e. corrupt or unavailable data) or complete, and it may be short lived or a long lasting event if several critical components are inoperable.

To reduce the probability of communication outages, special architecture and operations are required for the whole communication chain (from sensors collecting and emitting measures and orders to systems transmitting, receiving, storing and manipulating data, commands and communications).

4.3.5.2. Damage to the grid control centre

One hazard that could have very serious consequences is any event that could cause the grid control centre to be made inoperable. This might be the result of a natural event (e.g. a severe earthquake, a fire in the building, severe local flooding) or a malicious or terrorist attack. The integrity of the grid control centre and any backup control centre is the responsibility of the grid system operator, but the nuclear power plant operator and the nuclear regulator will need to be assured that suitable arrangements are in place to prevent a total collapse of the system if the control centre is rendered inoperable.

The risk can be reduced by ensuring that the grid control centre is not in an area where flooding is likely, and that the building has extremely good fire precautions and, if necessary, a construction that is very robust against earthquakes. The likelihood of malicious or terrorist attacks can be reduced by ensuring that the grid control centre is physically secure and its location is not public knowledge. It also needs to have computer security arrangements, such as those described in Refs [66–69].

Because of the seriousness of the possible loss of the grid control centre, it is advisable that there is a second or backup grid control centre at another location that has the capacity and capability to take over the main functions of control of the electrical system in an emergency. Some Member States already have a system where there is a single national grid control centre with responsibility for control of the highest voltage transmission system and the large power stations, but with several regional grid control centres that control lower voltage networks. With this arrangement, one of the regional centres could be modified to be able to act as a temporary national grid control centre when necessary.

For a backup grid control centre to be effective, it needs to have security arrangements that are similar to those of the main grid control centre. The members of staff there need to be well trained and have frequent and effective drills and exercises, including opportunities to practice the takeover of control. In some Member States, the staff are regularly exchanged between the main control centre and the backup control centre to ensure that they have the necessary experience.

4.4. DEFENCE IN DEPTH IN THE ELECTRICAL GRID

During a major grid incident, the various phenomena described in Section 4.2 may occur one after another or simultaneously with combined effects. System security and resilience are based on provisions of various kinds and levels and their implementation, and these need to be adapted to the dynamics of each phenomenon and act to prevent, detect and respond.

These provisions — which fall within the domain of material quality, engineering efficiency, organizational effectiveness, quality and competence in performance of activities, etc. — are recognized as the 'lines of defence'. The implementation of successive lines of defence is the main concept of defence in depth, in addition to the backup and defence actions. This principle is commonly applied in the field of nuclear safety (see Ref. [72]), as well as in many complex industrial systems for which a high level of safety is required. As a good (and necessary) practice, each of these areas are well defined with clear and concise entry and exit conditions and are provided with primary and backup strategies, plans, tactics, actions and instructions.

Accordingly, the defence in depth of electrical grids is based on the coherent integration of successive lines of defence to provide avoidance or control of the main phenomena and their ensuing stages that may lead to a total system collapse. Another role of defence in depth is to allow the recovery and restoration of grid security and functions in a timely manner.

The defence lines primarily relate to the following three areas:

(a) Prevention and preparedness;
(b) Monitoring, detection and action;
(c) Ultimate measures for recovery and restoration.

4.4.1. Prevention and preparedness

First, the defence lines for this area are designed to avoid the initiation of the phenomena. Actions are taken in normal operation state to do the following:

— Enforce the level of reliability, availability and performance of the components to function properly to ensure that the number of initiating events will be minimized. This is the primary reason for conducting preventive maintenance on the various components.
— Continuously ensure the availability of vital functions, even in the case of equipment failures. This is achieved by physical and functional redundancy, independence and diversity (e.g. redundant/diverse protection schemes for high voltage lines).
— Ensure proper performance of all activities by assessing risk for the power system by implementing quality assurance rules. This administrative task aims to improve settings, maintenance and operation of systems and components important for system security.

Second, the provision of operational margins improves system robustness in order to cope with faults or hazards that may weaken the system. Operational margins are security margins expressed in terms of operational means, allowing operators to bring the system back within its normal operating domain after an incident. Some examples of these margins are as follows:

— Active power margins are necessary to recover frequency after an unbalanced generation–demand event.
— Reactive power margins are necessary to recover voltage after an unbalanced generation–demand event or after a change in dispatch flows.
— Alternative options for power dispatch bring operational margins, allowing operators to reroute power to recover normal load flow, voltage or short circuits in substations.

These margins are variable in quantity and in means, depending on the grid status. All margins available without excessive costs are used to minimize the consequences of an incident, even if the consequences are acceptable.

It should be noted that reserves are particular operational margins used for covering scheduled operating schemes and listed contingencies (i.e. anticipated faults). For instance, active power reserves are used for frequency containment and to restore frequency, energy reserves for restoration of power reserves, reactive power to restore voltage plans and transmission reserves to give time for redispatch operation.

4.4.2. Monitoring, detection and protection

This area includes all actions, automatic (e.g. automatic frequency control) and manual (e.g. dispatcher actions) that detect variations in some key parameters of the system, such as voltages at all major nodes, power flows on all major circuits and system frequency. These variations can trigger appropriate corrective actions to ensure the security of the system. Typically, these actions are in response to the deviation from the normal operating state when the system has become more vulnerable to further incidents owing to the reduced security margins (i.e. it is in an alert state).

The objective of monitoring is to track performance characteristics in order to detect and predict degeneration, incidents and/or malfunctions. The objective of the automatic and manual protections is to take actions when incidents and malfunctions occur. Together, they prevent the system and its parts from further degradation or from becoming uncontrollable by exceeding their functional limits (as considered and incorporated in the design of the equipment and the system) and violating security limits.

4.4.3. Ultimate measures

The actions during an emergency state are aimed mainly at controlling the consequences of the phenomena described in Section 4.2 to avoid a total collapse of the network. If the system is no longer intact despite the ultimate measures implemented during the emergency state, further actions are taken to place the system in a situation facilitating its recovery or restoration. Some of these actions (i.e. ultimate measures) are automatic (e.g. automatic disconnection of demand in stages as the system frequency falls) and some are initiated manually by the grid system operator (e.g. instructions to disconnect blocks of consumer demand, to trip generating units or to split networks) as the system is outside its secure state.

4.4.4. Defence plan and plans for backup

The sequential lines of defence with matching corrective actions need to be implemented as dictated by the urgency of the situation and the degree of weakness of the system. This justifies the radical nature of the measures taken, sometimes at the cost of some deterioration in the provision of quality for a limited number of electricity consumers. The adopted philosophy — particularly for extreme situations where actions are taken as a last resort — is to intentionally isolate selected parts of the grid in order to preserve the integrity of the remaining system. These corrective actions can be grouped into two levels and are implemented in different timescales.

The first level includes actions to contain an event whose dynamics are still within the capabilities of human intervention (i.e. diagnosis, decision making and action on the system). These are protective actions that are part of 'monitoring and protection' and 'ultimate measures'. They include actions ensuring the supply–demand balance during changes to the generation schedule (e.g. rapid transition to maximum power, rapid electricity consumer shedding, emergency remote load shedding, interconnection tripping) and actions to maintain the voltage level (e.g. connecting reactive load groups, operating on-load tap changers). To ensure speed of execution, these actions need to be predefined. The instructions to carry out these actions may be issued globally on an area or to a given section of the grid.

The second level combines corrective actions to counter phenomena whose rapidity and evolution exclude the possibility of human intervention, and only automatic devices can effectively ensure the necessary remedial actions. The actions all fall within the domain of 'ultimate measures'. The defence plan can include the following:

— Automatic separation of areas that have lost synchronism;
— Automatic load shedding on low frequency;
— Automatic blockage of on-load tap changers for high and medium voltage transformers on voltage drop;
— Request for immediate power increase or decrease;
— Request for reactive power adjustment.

4.5. RECOVERY AFTER MAJOR BLACKOUTS

This set of safeguard actions and the defence plan are completed by a restoration plan, with the objective of facilitating a controlled, prioritized and timely restoration of power supply to the affected users, paying special attention to essential consumers and areas.

Generally, the grid system operator needs to establish a restoration plan in coordination and consultation with significant grid users, as well as the grid and nuclear regulatory authorities, and, if applicable, the neighbouring grid system operators and other system operators sharing the synchronous area (such as the distribution network operators).

4.5.1. Establishing the restoration plan

The first step in the restoration plan is to identify the relevant parties, particularly those who are essential grid users, and their applicable transmission and distribution system operators. In addition to the listing and classification of relevant parties, the restoration provisions that require agreements to be included in the plan are the following:

(a) The conditions under which the grid system operator activates the plan.
(b) Restoration plan instructions to be issued by the grid system operator and their guidelines, typical sequence and prerequisites.
(c) Measures that require advance and/or real time consultation or coordination with relevant parties (i.e. the grid system operator, significant grid users and the applicable distribution system operators) and their sequence and required completion times.
(d) Terms and conditions for the disconnection and re-energization of high priority, significant grid users.
(e) Specific restoration procedures, such as re-energization, frequency management, resynchronization, with technical and administrative instructions based on indications and symptoms.
(f) Identification and listing of time critical processes followed by the grid system operator and high priority, significant grid users.
(g) Essential and special elements of the system restoration process and procedures, such as:
    (i) Substations;
    (ii) Power sources in the grid system operator's control area that are used for the re-energization strategy, including their number, type and role (i.e. black start capability, quick resynchronization capability through house load operation and island operation capability);
    (iii) Dedicated substations and power sources for a specific, significant grid user.

The critical information — such as updates on the status of the restoration, the operational limits and their trends, the current capabilities for activation, the increase of generation and its possible times — needs to be made available to the grid system operator by the significant grid users.

It is important for the restoration plan to take into account the behaviour and capabilities of load and generation, the specific needs of the high priority, significant grid users and the characteristics of the network (including the underlying distribution networks). It needs to be ensured that any provision or measure that is prescribed in the restoration plan does not lead the interconnected transmission systems into an emergency or blackout state (i.e. not making a dire situation worse).

It is also a good practice to have a graded and reasonable approach for an effective restoration plan with the intention of minimizing the impact on system users while considering economic feasibility and efficiency. Details on restoration plans can be found in Ref. [73]. It is also very important not to dilute the 'absolutely necessary' actions and measures with 'good to have' actions and measures in the plan.

4.5.2. Restoration of power to the nuclear power plant

The principles of and processes for power restoration related to nuclear power plants, and the classification of nuclear power plants as significant grid users or high priority significant grid users, are based on the following:

(a) Priority ranking to re-energize: After a blackout, those installations (e.g. remote grid control centres, grid system operator's critical installations, auxiliary services of power plants) that, if unavailable, could cause a significant social or environmental impact, or could compromise the continuation of the successive switching sequences, or which are of importance for guaranteeing stability of the grid, need to be given priority to be resupplied with a time delay that is compatible with their autonomy.

(b) The design and operational characteristics of the nuclear power plant, such as:

  (i)   Amount of off-site power: The amount of off-site power required to repower essential nuclear power plant auxiliaries, including the associated short circuit power, should be clearly defined.

  (ii) House load operation capability: Some nuclear power plant designs allow the plant to disconnect from the grid and provide power just to the house load (i.e. house load operation), noting that a nuclear power plant with this capability could support grid restoration by load pick-up for adjacent local sections of the grid. This is based on the assumption that the automatic voltage regulator and governor are able to balance the reactive and active power surplus or deficit after load pick-up, and it is permitted by the nuclear regulations.

  (iii) Islanded operation capability: Some nuclear power plant designs allow a plant to supply some local demand on the local grid while disconnected from the main grid. A nuclear power plant can bring stability into an island and during load pick-up steps in the case of re-energizing the current island. This requires that the automatic voltage regulator and governor can balance the reactive and active power surplus or deficit after load pick-up or load shedding in small islands.

Whatever the reasons for a power disconnection (i.e. LOOP) at a nuclear power plant grid connection point are, it is the responsibility of the grid system operator to restore stable power to the nuclear power plant as soon as practical. Clear and unambiguous communications between the nuclear power plant and grid system operators are an absolute necessity.

5. ASSESSMENT OF THE RELIABILITY OF THE GRID CONNECTION TO THE NUCLEAR POWER PLANT

The reliability of the grid connection to a nuclear power plant can be analysed in two parts:

(a) 'Grid connection studies,' which include the functions connecting the nuclear power plant to the grid and the interactions between the nuclear power plant and the grid in operation. Similar studies are needed for the connection of any large new power plant.

(b) 'Off-site reliability studies,' which include the infrastructures of the grid–nuclear power plant interface and the off-site power sources. These studies are required for a nuclear power plant but may not be required for other large power plants.

These analyses can generally be performed using one of several commercially available software packages for network analysis, as described in more detail below.

To carry out these analyses, the nuclear power plant developer needs to provide data and specifications for the systems and components of the nuclear power plant, such as the parameters of the turbogenerator and transformers and the preferred settings for their controls. The grid system operator and transmission system owner need to provide the electrical specifications and parameters for the electrical grid, such as the impedances of overhead lines, circuit breaker ratings, fault clearance times, short circuit level at the connection point and planned outage durations. The technical criteria used for the assessments of the simulations need to be agreed by both parties.

The results of the analyses may be reviewed by the nuclear regulatory body and may be integrated in the safety analysis report (see IAEA Safety Standards Series No. SSG-61, Format and Content of the Safety Analysis Report for Nuclear Power Plants [74]). In some Member States, the nuclear regulator imposes a list of simulations to be integrated in the safety analysis report.

5.1. GRID CONNECTION STUDIES

At an early stage of each nuclear power plant project development, grid connection issues need to be evaluated. It is necessary to carry out specific studies to define the optimal technical and economic solutions for the grid connection and the design of the connecting (main and backup) substations.

The grid connection studies need to implement the nuclear safety requirements related to off-site power supply. Because of this, it is necessary to perform a grid reliability assessment at the locations where it is proposed to connect the nuclear power plant with the grid.

Power studies and associated static and dynamic grid modelling are used for performing grid connection studies and address the following areas:

— The grid capacity to export full nuclear power plant power to load centres in normal and in grid contingency situations (e.g. load flows in transmission circuits, voltage at grid nodes, short circuit currents in grid substations);
— The stability of the nuclear power plant and of the grid in any operating mode of the nuclear power plant as well as their behaviour following any possible nuclear power plant transient (e.g. synchronization, active or reactive power steps, trip to house load, load rejection) or grid transient (e.g. power oscillations, voltage and frequency deviations, activation of line protections).

All the grid components that may have an impact on nuclear power plant capacity or on stability (e.g. HVDC connectors, generation or demand facilities connected by power electronics systems, static volt-ampere reactive compensators, special protection schemes) need to be included in the power studies.

5.1.1. Capacity of the grid

Three types of study address the capacity of the grid:

(a)   Load flow studies;
(b)   Steady state security studies;
(c)   Short circuit studies.

The load flow study calculates the magnitude and the phase angle of voltages in each network node and the currents through each grid element for a certain network topology, generation pattern and demand distribution. Power flows through each transmission line and transformer can be compared with their rated power, so possible overloads or congestions in a grid may be detected.

The load flow study may also be used for determining the range of reactive capability that is required of the generator, and hence the necessary parameters of the unit transformer (i.e. rating, turns ration, tap range and likely tap setting).

The steady state security study is an extension to the load flow study, verifying that the system can be operated in accordance with the $N-k$ criterion as described in the Glossary and in Section 4.1. A steady state load flow calculation is carried out both before and after removing one or more circuit elements (e.g. overhead line, underground cable, transformer) from service. If removing those circuit elements causes any voltages to be outside the desired range, or causes any circuit elements to be overloaded, then the security criterion is not met.

For evaluating the nuclear power plant grid connection, it is expected that at least the $N-1$ security criterion can be fulfilled for the main transmission lines near the nuclear power plant, so that there will be no cascade tripping of transmission circuits that may jeopardize connections to the nuclear power plant. For the nuclear power plant grid connection design, additional $N-2$ or $N-1-1$ criteria may be studied by observing the effect of simultaneous forced outages of two circuits or the forced outage of one circuit while another one is on planned outage for maintenance. One conclusion from such studies may be that certain planned transmission circuit outages for maintenance may be permitted only during reactor outages or if the reactor is operated at reduced power.

Similar criteria can be used in the design for the off-site power supply to the nuclear power plant (i.e. the supplies to the standby transformers). The forced outage of one off-site power supply element (i.e. line, cable, transformer or breaker) needs to not cause the interruption of the supply to the standby transformer and hence the safety buses.

Violations of the security criteria detected in steady state security studies may be removed on a permanent basis by suitable grid reinforcements or, if that is not possible, by suitable outage planning and dispatching actions, depending on their efficiency, problem mitigation speed and associated costs.

A short circuit study identifies whether the short circuit currents at substations nearby, due to the new generator, may exceed the ratings of substation equipment, particularly of circuit breakers. If they do, then it is necessary to upgrade such equipment with a higher short circuit rating, or take other measures to limit the fault current, to ensure that the equipment will operate correctly under fault conditions. There are several standards on the short circuit calculations in the electricity grids, which may give different computational results. It is important that the appropriate standard be used to match the definitions for the fault ratings of equipment.

5.1.2. Stability of the nuclear power plant and the grid

Grid stability is defined as "the ability of an electric power system, for a given initial operating condition, to regain a state of operating equilibrium after being subjected to a physical disturbance, with most system variables bounded so that practically the entire system remains intact" [75]. This ability can be understood as robustness, and a more robust system is expected to be more reliable. For a new nuclear power plant, it is important to demonstrate that the connection arrangement chosen will allow the grid system to be stable with the nuclear power plant in operation.

There are several types of stability problem in a power system, namely, angle stability, voltage stability and frequency stability. These are described in detail in standard textbooks such as Ref. [76]. These stability problems are identified using dynamic simulations and computations evaluating the consequences for grid stability in a nuclear power plant operation. In the case of stability issues (e.g. undamped oscillations, large voltage or frequency deviations, loss of synchronism), the remedial actions (on the grid side or the nuclear power plant side) are checked again and iterative studies are run until the stability issues are fixed.

Usually, a power system stability study includes the following four steps:

(1) Make modelling assumptions and formulate models that are appropriate for the timescales and phenomena under study.
(2) Select appropriate criteria to assess the outcomes of the study.
(3) Analyse and/or simulate the power system to determine the behaviour of relevant technical parameters, typically using a scenario of events.
(4) Review the results considering assumptions, compare them with criteria and engineering experience, and repeat the stability study if necessary.

Simulations for rotor angle stability are required to demonstrate that the generator will remain synchronized and that oscillations of the rotor angle will be sufficiently damped following a large disturbance (transient stability) or small disturbance (small signal stability) on the grid. For the transient stability and small signal stability calculations, detailed mathematical models of the generators, their excitation systems, turbine governor systems and related protection systems are necessary. The events typically analysed are as follows:

— A three phase short circuit close to the nuclear power plant (in the substation near the nuclear power plant or on a transmission line connected to it);

— Asymmetrical faults such as a single phase short circuit close to the nuclear power plant substation;
— A small change of nearby load or a step change of the reference value in the generator's excitation system.

It is important that none of these events cause stability problems. It is expected that the generator can ride through nearby short circuit faults that are cleared by normal operation of the electrical protection. It is also expected that the generator can withstand any kind of small disturbance, meaning that it remains in synchronous operation and any oscillations following the disturbances are small and well damped.

The simulations for voltage stability check the ability of the power system to maintain steady voltages on all buses during voltage and active or reactive power transients. The following list indicates the nuclear power plant functionalities to be integrated in the simulations:

(a)   It is expected that the reactive capability of the generator (i.e. the voltage stability curve) and the controls of the exciter are sized and tuned to withstand any kind of small or large disturbance, as specified in the grid code, meaning that the nuclear power plant remains in synchronous operation and any oscillations following the disturbances are small and well damped.
(b)   The voltage regulation is enabled to ensure that load connections causing dips of voltage are automatically regulated. Nuclear power plant protections need to be stabilized against in-rush currents if they do not exceed the agreed upon protection schemes.
(c)   To ensure that adequate synchronizing power is maintained when the nuclear power plant is subjected to a large voltage disturbance, the exciter whose output is varied by the automatic voltage regulator needs to be capable of providing its achievable upper and lower limit ceiling voltages to the generator in a time not exceeding that specified by the grid system operator.
(d)   The model of the excitation system may be required to include a power system stabilizer to prevent or attenuate power oscillations, with the following functions:
    (i)   Limitation of the bandwidth of the output signal;
    (ii)   Limitation of underexcitation;
    (iii) Limitation of the maximum overshoot in response to a step injection that operates the underexcitation limiter;
    (iv)   Limitation of overexcitation.
(e)   The modelling needs to demonstrate that a sudden load reduction does not lead to undamped oscillations in active or reactive power.
(f)   The generation of maximum leading and lagging reactive power capability is included in the simulation in accordance with the generator specifications.

The simulations for frequency stability verify the ability of the nuclear power plant to stabilize the load and the speed of the generator as well as the characteristics of the frequency control in compliance with the nuclear power plant specifications and grid requirements. The following list indicates the nuclear power plant functionalities to be integrated in the simulations:

— Modulation of active power at high and low frequency (or in the full frequency range if the nuclear power plant will participate in frequency control) carried out by means of frequency steps and ramps reaching the minimum regulating level, considering the droop settings and the dead band;
— Capability to ride through faults in accordance with the conditions set out in the grid code;
— For island mode operation, capability to stabilize the frequency and ability to reduce or increase the active power output from its previous operating point to any new operating point in the active and reactive power (P–Q) capability diagram (within the limits), without disconnection due to over- or underfrequency.

5.2. OFF-SITE POWER RELIABILITY STUDIES

5.2.1. Relevance of the assessment of the reliability of the off-site power system

Since the grid connection of a nuclear power plant is required to be designed to ensure reliable power supply of the plant during reactor startup, normal operation and shutdown, off-site reliability studies are performed for the off-site power system for each mode of plant operation to assess the reliability of different options of the design.

Failures and abnormalities in the off-site power system can result in the disconnection of the nuclear power plant and consequently an automatic or manual power reduction to house load or an automatic trip or manual shutdown of the nuclear reactor. This results in a situation where the plant relies on on-site power supplies until the preferred power supply (from the off-site electrical grid) is re-established.

A LOOP is considered an initiating or a concurrent event with a predicted frequency and probability in the safety analysis of a nuclear power plant. Prevention, protection and mitigation measures are put in place in anticipation of such an event; for example, backup alternating current (AC) sources. Furthermore, since these engineered backup AC sources have a limited operational time, additional measures, such as supplemental alternative AC and DC power sources, may be needed to mitigate a complete loss of AC power (i.e. SBO) for a limited period that covers the time until the restoration of the off-site power supply.

To reduce the risk and consequences of LOOP and SBO events, they need to be rigorously analysed. Design provisions and administrative controls are established as described in Section 7. In order to analyse, identify and select appropriate design provisions, the probability of losing the off-site power from the electrical grid and the duration of LOOP conditions (i.e. grid power restoration time) need to be estimated.

5.2.2. Methodologies

Most methods for the determination of reliability indices of the off-site power system use approximation or simplification to reduce a complex and nonlinear system problem to a solvable level.

Grid reliability indices observe specific consequences in a grid (e.g. line overloading, bus voltage violation, loss of load, expected unserved energy, load curtailments) together with related probabilities. In the overall grid reliability assessment, every grid element (line and busbar) is defined with the average frequency and average duration of outage, which may be divided between forced and planned ones. Outages may be single, multiple independent, common cause multiple dependent, substation related or multiple dependent. The total probability of the observed contingency (e.g. that all lines connecting a nuclear power plant substation with the grid are out of operation) is calculated as a sum of the probabilities of multiple independent events and multiple dependent events (common cause and substation related).

For a nuclear power plant grid connection design and preferred power supply design, it is important to consider the probability that the nuclear power plant's house load may not be supplied from the grid because of different single, multiple independent and common cause multiple failures/outages. The general criterion is to design the nuclear power plant grid connection and the preferred power supply to minimize the probability of grid related and substation related LOOP events. The following list indicates some generally used methods:

(a) Statistical methods assess the frequency of LOOP events on the basis of historical data of event occurrence and the specific conditions as to when or where these events occurred. Although these methods can also calculate the event frequency simply using the number of events in the time period when data are available (i.e. in the period of time considered in the assessment), they allow the calculation of the LOOP event frequency for the times when an anticipated and specific risk or vulnerability to a certain cause or consequence exists. The recorded LOOP event occurrences, therefore, are divided into different groups (e.g. causes, location, personnel involved) to calculate the frequency specifically covering hazards, risks and vulnerabilities.

(b)   Probabilistic methods make it possible to determine the probability of unwanted events (e.g. LOOP at the standby transformer) on the electrical grid and an indication of the level of severity. This capability provides a number of key benefits, including the identification of the most critical generator or grid failures and the determination of the most effective mitigation actions.

(c)   Fault tree methods are based on the assessment of the consequences of contingencies (grid elements and generators) and require long computational times and the investigation of a huge number of system and component outage cases. Therefore, these methods use simplified, equivalent models of the systems and look at reasonably credible scenarios.

5.2.3. Data

If analysis is started from the component level (e.g. circuit breakers, disconnect switches, lines, transformers), input data for failures for different voltages and with a relevant (i.e. adequately long) time period are needed. Input data for loads and generation at each substation are needed in the fault tree analysis approach.

The number of failures and their duration are essential inputs for the statistical analysis of the grid reliability. Information on the voltages and locations of the failures can provide additional insights on the analysis. For the assessment of the LOOP event frequency, classification of the events by cause, location and operational mode of the nuclear power plant is needed.

It is expected that the input data for these studies will be collected, some of which are provided by the grid system operator. Input data from equipment vendors, power plant operators and other relevant entities will also be used in the analyses. As reliable and confirmed inputs are essential for the correctness of the results, assumptions and expert judgement on the parameter values need to be minimized and, if possible, avoided. If the use of such assumed/judged input values is unavoidable, the basis for the assumptions and a sound justification for the expert judgement are to be provided (including the means and the time of a later verification of such contingencies).

5.3. EXPECTED RESULTS

The obtained results are expected to be compiled in a report including the following:

— The overview of the system, including the neighbourhood, the configuration of the transmission grid and the nuclear power plant's connecting substation;
— The reliability of the feeder circuits (e.g. method, analysis, restoration time, history of circuit trips);
— The reliability of the busbar and the protection systems (same as above);
— The concerns that are included in the analysis (i.e. weather related, resources and outages);
— The rare events that are not included in the analysis (e.g. historical records, aircraft crashes, geographical spread, earthquakes, floods, storms);
— The power system disturbances (i.e. voltage, frequency, power quality and planned system conditions);
— The estimation of the probability of a LOOP at the nuclear power plant site;
— The references for the sources of data and tools used.

Table 2 shows typical outputs from the discussed reliability studies.

TABLE 2. SUMMARY OF TYPICAL OFF-SITE POWER ASSESSMENT OUTPUT FROM DIFFERENT METHODS

Output	Method		
	Statistical assessment	Probabilistic assessment	Fault tree analysis
Frequency and duration of power grid faults	Data provided by the grid system operator (e.g. electricity consumers interrupted, duration of interruption, number of momentary interruptions, connected kVA load served and/or interrupted)	Input: inherent reliability, mean time to repair (in hours), failure rate (failures per year), single line diagram	Loss of load (main and auxiliary) Input: inherent reliability, mean time to repair (in hours), failure rate (failures per year), single line diagram
LOOP frequency	Frequency and duration of LOOP	Calculation of LOOP probability	Trip to islanding failure
Mean time to restore off-site power	Data required from the grid system operator (e.g. time taken to identify the problem, repair it and restore power; time taken to mobilize repair crew; availability of spare parts)	Evaluation of the probability to exceed a given value	
Reliability of the substation	Failure rates for main components Experience feedback from the grid system operator and industry associations (e.g. CIGRE)	Comparison between vendor data sheets and data collected during operation	Identification of main contributing components

Note:
— The frequency and duration of the grid faults in the whole system or at a specific point are expected to be assessed in the case of a statistical analysis.
— The LOOP frequencies, due to different causes and during any modes of plant operation, are established in the LOOP frequency assessment.
— The probability of restoration of power after a given time interval can be assessed if input data are provided.
— The reliability of the power system in all load points (i.e. substations) and overall is obtained from the fault tree analysis method.

5.4. GRID CONNECTION RELIABILITY CONCLUSIONS

To determine the possibilities of hosting a nuclear power plant on an electrical network and to identify the consequences of its inclusion in the electrical system, the grid system operator and the nuclear power plant developer study the impact of the installation by evaluating the conditions for compliance with a certain number of technical criteria. These studies make it possible to determine whether the integration of the new nuclear power plant, depending on its connection point, is subject to any restrictions on use or network reinforcements.

The grid system operator also studies the feasibility and the technical conditions for making the connection (e.g. constraints of size and technical compatibility of the connection to the grid substations;

feasibility and environmental acceptability of the connection link). The depth of the analysis is adapted to the nature of the connection request to best meet the needs of the nuclear power plant project.

The important components for such evaluations are the following:

— The data, tools and methodologies used for the power studies and for the assessment of the reliability of the grid connections;
— The skills of the engineering teams performing the studies;
— A trusting and open collaboration between the nuclear power plant developer and the grid system operator teams to reach a connection agreement in compliance with the nuclear safety rules and with the grid code.

6. DESIGN OF MAJOR ELECTRICAL COMPONENTS RELEVANT TO RELIABILITY

This section discusses the main design provisions and technical options for the reliability of electrical grid components that significantly influence a nuclear power plant's design and performance.

6.1. SUBSTATION (SWITCHYARD)

The objective for the substation design is to comply with system requirements and provide maximum reliability and continuity of service with reasonable investment costs. The design may also provide for further expansion. There needs to be sufficient flexibility to allow for maintenance outages of lines, circuit breakers and switches with no service interruptions or hazard to personnel.

The connection of the substation to the electrical grid uses several overhead lines or underground cables, with separate routes connecting to separate grid substations in accordance with the guidance provided in SSG-34 [4]. The reliability of this arrangement is verified by the studies described in Section 5.

Many factors influence the selection of the design of a substation for a given application. A variety of substation configurations are used in Member States. Generally, each grid system operator and transmission system owner has a preference for particular configurations, based on previous experience and influenced partly by the ownership arrangements for substations. In some Member States, the entire substation at the nuclear power plant is owned by the transmission system owner. In others, the substation belongs to the nuclear power plant, with the transmission system owner being responsible only for the outgoing circuits. In yet others, the transmission system owner owns the main part of the substation, but the nuclear power plant owns and operates the circuit breakers that connect it to the substation. It is helpful in operation if the substation configuration at the nuclear power plant is similar to those of the other substations operated by the same grid system operator and transmission system owner.

There may also be differences in design, depending on local conditions. There may be a preference for indoor substations because of local weather conditions (e.g. in the United Kingdom, for substations near the coast). With evolving technology, there has been a trend towards fully gas insulated substations, instead of air insulated ones. They can be more compact, although they are more expensive.

In the case where the substation is designed by the organization responsible for the nuclear power plant (i.e. the technology vendor or architect engineer), it is necessary for the grid system operator and transmission system owner to agree on the design of the substation and on its compatibility with the design of the transmission system lines connecting to it. It is also necessary to agree on the protection systems; these are generally the responsibility of the transmission system owner, depending on the off-site power boundary agreement. Several factors need to be considered in the selection of bus layouts and switching arrangements for a substation to meet system and station requirements.

For reliability, the substation design needs to prevent any total shutdown of the substation caused by breaker failure or bus faults and needs to permit rapid restoration of service after a fault occurs. A planned arrangement of lines with sources connected opposite to loads improves reliability. It is good practice to have the layout permit future additions and extensions without interrupting service.

It is an industrial safety requirement that all equipment within a substation be designed to survive the maximum short circuit current that may appear owing to nearby faults within a substation or on the transmission lines.

From the perspective of reliability of the grid–nuclear power plant interface, it is necessary to pay particular attention to the main bus connections in the substation(s) to which the nuclear power plant is connected; the substation controls and protections; the protections against severe weather; the special control systems; and the computerized substation control systems.

6.1.1. Main bus connections in the substation to which the nuclear power plant is connected

There are various connection schemes in Member States and each grid system operator and transmission system owner prefers one common type of scheme owing to the ease of maintainability while providing adequate reliability.

For the enhanced reliability of off-site power to a nuclear power plant, it is important to ensure that there is at least one double busbar and two independent breakers between the connection to the unit transformer and the connection to the standby transformer when designing the single line diagram of the substation to which the nuclear power plant is connected (see guidance in Ref. [6]). This is to ensure that a single fault to a section of the busbar or a fault to a circuit breaker (which usually affects two sections of the busbar) cannot simultaneously affect the connections to both the unit and the standby transformers.

The large number of possible substation geometries are described in various standard engineering texts such as Ref. [77]. Four arrangements that have been used at nuclear power plants are as follows:

(a) Ring bus, which is suitable for a single nuclear power plant (see fig. 5 in Ref. [6]);
(b) Double busbar, single breaker, where there are two separate busbars and each connected circuit can be connected to either busbar, but to only one at a time;
(c) Double busbar, double breaker, where there are two separate busbars and each connected circuit can be connected to both busbars simultaneously;
(d) Breaker and a half, where there are two separate busbars and each connected circuit can be connected to both busbars simultaneously, but the circuit breakers are shared with one or more other circuits.

6.1.2. Controls and protections

The choice of substation arrangement influences the complexity of the substation control and protection system and its reliability. Increased reliability and operational flexibility mean a higher initial investment and possibly higher maintenance costs. Therefore, the main design factors for the substation arrangements and for the choice of the protection technology with a direct impact on the reliability are as follows:

— Surety. The controls and protections are not meant to operate in an unexpected manner.
— Selectivity. The fault and its location are properly identified so that only the part that is concerned is disconnected/isolated in order to minimize unintended power outages.
— Agility. The controls and protections operate as rapidly as possible, which is equal to or better than the response time requirements.

— Simplicity. The controls and protections consist of the minimum number of active components and cable connections, since a design approach based on simplification improves reliability and reduces human interference as much as possible (see Ref. [78]).

6.1.3. Provision against severe weather

The effects of weather and other hazards on the reliability of the electrical grid are summarized in Section 4. The hazards that are important for a substation are as follows:

— Heat, drought and dust;
— Flood;
— Rain and humidity;
— Cold, snow and ice;
— Wind.

Reference [79] describes in detail the effect of these severe conditions and the measures to mitigate or protect the substations from such effects. Sometimes, the effects of two or more simultaneous weather related events need to be considered in the design (e.g. ice combined with high winds).

6.1.4. Computerized substation control systems

Computerized substation control systems are the systems that perform and integrate the basic operating functions of a substation, including the following:

— Equipment and parameter monitoring;
— Metering;
— Fault recording;
— Device control;
— Communication;
— Substation supervision.

The architecture of a digital substation includes three levels:

(a) The station control area level. A digital substation is based on a communication architecture, whereby real time operational measurements are polled from the site level. These data are obtained using sensors embedded within the site system. Measurement data are communicated to devices that act on those measurements by means of a 'process bus'. Smart devices and systems within the substation may also immediately process these operational data.

(b) The protection and control level. Between the process bus and the station bus are devices historically identified as 'secondary equipment'. In the digital substation, these devices are intelligent electronic devices, such as protection systems that communicate with a process bus, other peer devices in the same bay or in other bays, and the digital control system via the station bus.

(c) The primary equipment process level. This level captures and consolidates data from meters, breakers, sensors, etc.

The reliability of such computerized system is specified, for example, in IEC Standard 61850-3 [80] as follows:

"The substation shall continue to be operable, according to the 'graceful degradation principle', if any substation automation system communications component fails. There should be no single point of failure that will cause the substation to be inoperable. Adequate local monitoring and control shall be maintained. A failure of any component should not result in an undetected loss of functions nor multiple and cascading component failures."

Modern digital substation control systems are highly configurable, which means that stringent data description and configuration management are necessary. Digital and point to point cable solutions are often used for line protections in substations that connect a nuclear power plant to the grid. These connections have to provide a flexible, modular and reliable solution to allocate protection and communication functions between digital and point to point systems.

Specific attention needs to be paid to the design and reliability of the communication systems used for carrying protection signals, as the operating experience in some Member States has highlighted occasional undesirable system behaviours.[15]

6.2. MAJOR PLANT COMPONENT DESIGN FOR GRID RELIABILITY

Some of the major components that belong to a nuclear power plant have design and operational characteristics that affect the reliability of the electrical grid that they are connected to. Furthermore, their proper and adequate design and operation rely on the reliability of the grid system. Therefore, at the design stage of these components, the grid system's needs and requirements need to be shared and discussed among the designer responsible for the nuclear power plant (i.e. technology vendor and/or architect engineer), the nuclear power plant project owner, the grid system operator and the transmission system owner. These needs and requirements are considered and incorporated in the plant design and are controlled and maintained during plant operation.

The following sections provide an overview of the key design characteristics of some major, preferred power supply components of the plant in terms of their design and operational aspects in relation to the reliability and resilience of the electrical grid.

6.2.1. Main generator features

The main parameters and physical operation conditions of the generator (e.g. voltage, temperature, vibrations, reactive power gradient) are set according to the nuclear power plant's design and they are important for the electrical grid's reliability, as they are used particularly in designing the electrical protection of the system, in calculations of short circuit power and in modelling in dynamic and static grid simulation tools (see Section 5). For example, the grid system operator may specify the performance required from the generator excitation system (i.e. automatic voltage regulator) in order to ensure the transient and dynamic stability of the grid, such as a fast acting system with a high ceiling voltage. This requirement by the grid system operator necessitates that the nuclear power plant owner/operator or responsible reactor designer performs an analytical determination of transient stability issues and of the synchronous operation of the nuclear power plant unit(s) following different faults. For example, the maximum voltage that may occur after a loss of load will be affected by the speed of the generator excitation system and the ceiling voltage (see Ref. [6]). On the other hand, this maximum voltage has to be compatible with the allowable voltage range of the electrical supplies to the plant auxiliary equipment.

Generators are also equipped with engineered control systems tuning active power, reactive power and voltage at the generator terminals, as well as damping and stabilizing voltage and power fluctuations in support of the grid stability needs. These parameters and actions also determine the capability and

---

[15] In France, on 6 March 2004, four circuits connecting two plants (a total of eight reactor units) opened because of a malfunction of remote protection radio communication systems and reclosed in a few seconds. Following this event and after a series of iterations with the National Telecommunication Authority, the French grid system operator decided to install 9000 km of fibre optic cables, one quarter in a 400 kV network, another quarter in a 225 kV network and the remaining 4500 km in 90/63 kV networks (see Ref. [81]).

capacity for providing active and reactive power to the electrical grid, which in turn contribute to its reliability.

As such, some main generator parameters that are important for electrical grid reliability — and are also required and monitored (and collected) by the grid system operator — include the following:

— Reactive power capability, as it is related to the voltage regulation ability and determines the control area for voltage at the substation to which the nuclear power plant is connected.
— Active and reactive power controls and limitations, since they define the capacity of the generator to compensate different power and voltage transients, and their inadequate settings may reduce the reactive power capacity.
— Stability control (i.e. the electrical angle between the generator and the grid), owing to its importance for maintaining the dynamic stability of the grid, as well as its importance in a potential nuclear power plant trip resulting from the amplification of small voltage deviations by inadequate controls.
— Settings of generator protections linked to grid reliability (e.g. loss of synchronism, stator voltage, frequency, unbalanced currents, overload of rotor current), since inadequate protection can result in spurious actuation and an undesired trip sequence, causing a disconnection or a failure to disconnect the generator.

6.2.2. Unit transformer

It is a common practice to design the unit transformer of a nuclear power plant and its connection to the grid using the same engineering standards that are applied in any connection of a non-nuclear power generator with a large power transformer. As explained in Ref. [6], the design and operation of unit transformers and tap changers may vary from one Member State to another, depending on the specific requirements of the grid system operator.

In some Member States, the unit transformer has an off-load tap changer, so that in normal operation, the transformer is operated at fixed tap, and the reactive power, and hence the grid voltage, is controlled by varying the set point for the generator terminal voltage. In this case, it may be necessary for the auxiliary transformer to have an on-load tap changer to control the range of voltage supplied to the auxiliary plant.

In other Member States, it is normal practice for the unit transformer to have an on-load tap changer, and the generator to be operated at fixed terminal voltage. In this case the reactive power, and hence grid voltage, are adjusted by carrying out tap changes, and the auxiliary transformer is operated at a fixed tap position.

Repairing or replacing a unit transformer could be a significant issue and a challenge for timely system resilience and availability, owing to the unavailability of spare parts and the elaborate and lengthy manufacturing and transportation processes for large power transformers. Therefore, the nuclear power plant owner/operator needs to establish adequate processes for the procurement of new transformers or refurbishment services, the purchase and maintenance of supporting components and the inventory control of spare parts to avoid a long term service interruption and considerable economic loss for both the nuclear power plant and the grid owners/operators.

Furthermore, during plant operation, a high level of transformer availability and reliability needs to be maintained by adopting adequate provisions. There are several guidance publications for maintaining and improving large power transformers. For example, the guidance from the World Association of Nuclear Operators (WANO) [82], provided to plant operating organizations, recommends the following:

— Implementation of effective monitoring;
— Provision of life cycle management to preclude in-service failures;
— Identification and mitigation of single point vulnerabilities;
— Development and use of high quality instructions, procedures and work orders.

6.2.3. Auxiliary transformer

Different nuclear power plant design and operation concepts use different arrangements for supplying the house load, depending on the technology, off-site power source, grid characteristics or market conditions. For example, it may be more economical to use an auxiliary transformer for the house load during normal operation rather than buying power from the grid, or it may be more reliable to supply approximately half of the auxiliary equipment from the auxiliary transformers and half from the grid via the standby transformer. As such, the power to the auxiliary equipment (i.e. house load) of the plant (which is several tens of megawatts) can be supplied in the following ways:

— By extraction from the main generator (or unit transformer) output during normal operation;
— Directly from the electrical grid when the electricity from the main generator output is not available.

Normally, standard industrial grade transformers are utilized for the auxiliary transformers; however, nuclear power plants typically have at least two auxiliary transformers in the on-site power system in order to provide redundancy. An alternative approach is to use multiple windings on the secondary side of the transformer. Further, their rating has to be at the level of the house load of the plant to power all relevant SSCs of the electrical power system of the plant.

Depending on the nuclear power plant's technology and size, the auxiliary transformers may be connected to the generator terminals, as shown in Fig. 6. This is the most common arrangement. Alternatively, the auxiliary transformers could be connected directly to the grid voltage at the local substation, as shown, for example, in fig. 6 of Ref. [6]. This latter arrangement would allow maintenance of the unit transformer while still providing two independent supplies (via the auxiliary and standby transformers) to the plant auxiliary equipment.

Like the unit transformer design features for the voltage stability, the auxiliary transformer may have an on-load tap changer or an off-load tap changer. The correct tap position allows the voltage to the electrical auxiliaries to be maintained within the necessary range for the full range of variation of the grid voltage.

6.2.4. Standby transformer

The safety case of the nuclear power plant may require the nuclear power plant to reduce power and to reach a safe shutdown state within a defined time if the standby transformer is not available. Hence, the availability of the standby transformer is important to ensure that the nuclear power plant is available to generate electricity.

In most nuclear power plants, these transformers are typically in standby mode (i.e. they are energized from the grid but do not carry a load) until the supply from the auxiliary transformer is not available. Hence, the functionality and operability of standby transformers need to be checked periodically and the availability and quality of electrical power from the grid needs to be regularly confirmed. When the standby transformers are in service, they need to be able to supply the full auxiliary load of the nuclear power plant.

From the plant safety and reliability perspective, it is good practice in design that the connection of the standby transformer to the electrical grid is independent and separated from the connection of the unit transformer, as emphasized in Section 6.1.1. This can be accomplished by the connection of the standby transformer to a different substation (possibly at a different voltage) or within the same substation but with adequate separation from the connection to the unit transformer. In some Member States, the standby transformer is powered from an independent generating unit and substation.

During operation, the minimization of repair and replacement time with readily available spare parts is also critical for providing a high level of availability. Therefore, adequate processes need to be established and updated for the procurement of new transformers or refurbishment services, the purchase and maintenance of supporting components, and inventory control for spare parts.

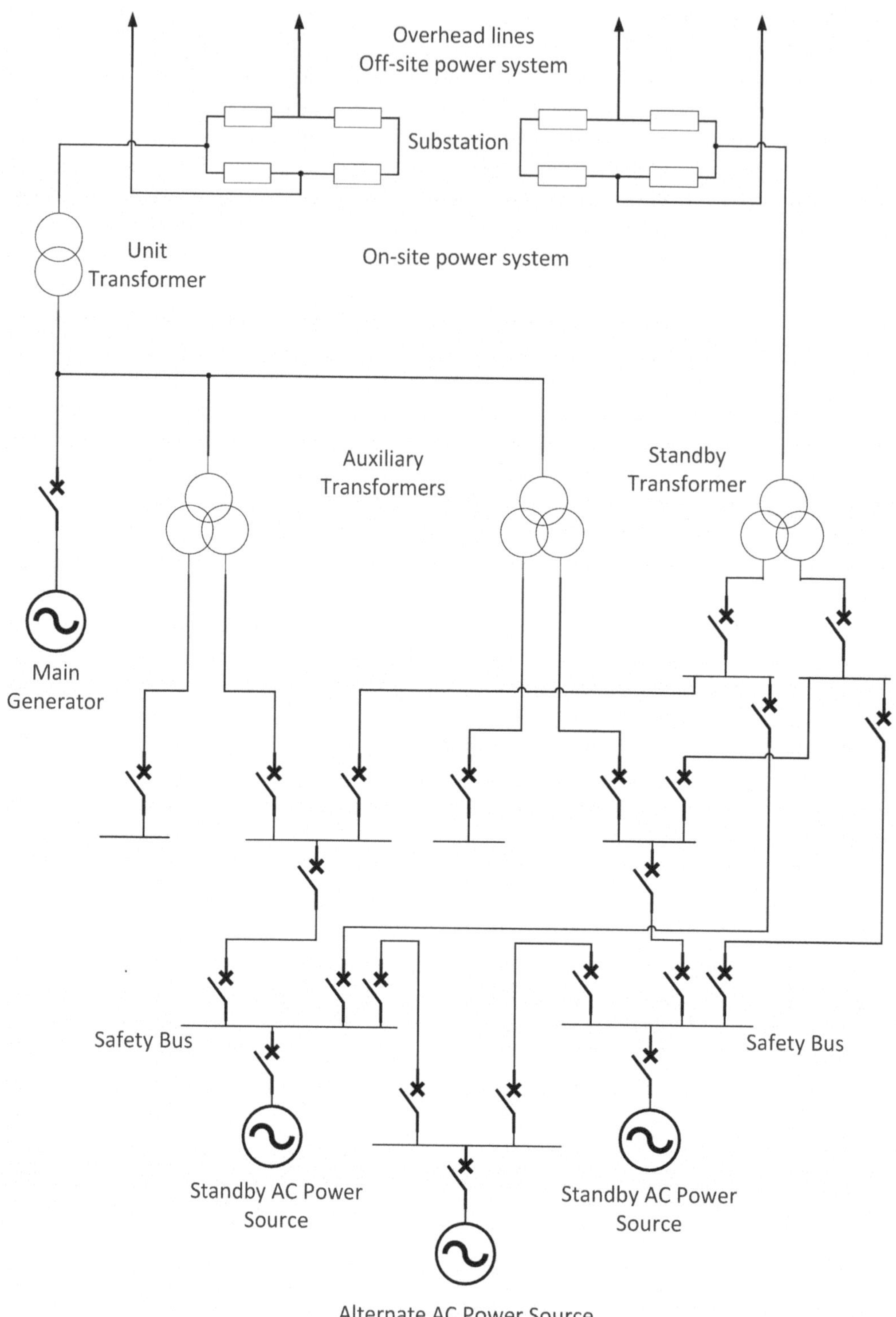

FIG. 6. Auxiliary transformers connected at the generator terminals. Adapted from fig. 1 of SSG-34 [4].

6.2.5. Provisions to reduce the probability of a common fault in connections to unit and standby transformers

Independent of the reliability of the substations to which these transformers are connected, a few provisions regarding the arrangements of the components are important for the overall reliability of the off-site power sources. These necessary provisions to reduce the probability of a common fault in connections to the main and standby transformers may include the following methods:

— If the main transformer and the standby transformer are connected to the same substation, the connections are separated by two or more independent circuit breakers.
— If the electrical lines connecting the unit transformer and the standby transformer to the substation are supported on towers, it is advisable that they are not on the same row of towers and that the distance between the two rows of towers is large enough to avoid a common failure of the two lines because of a tower falling.
— The earth wire cables in the substation need to be located so that they cannot fall on the connections to the standby transformer and the unit transformer and the associated sections of the busbar.
— The circuit breakers need to have sufficient physical protection to protect them from damage due to catastrophic failure of other circuit breakers.
— The measurement and protection systems of the unit transformer connection and of the standby transformer connection have to be fully independent and physically separated.

6.3. GRID SUBSTATION(S) AND TRANSMISSION LINES

The nuclear power plant off-site power system may also include other substations identified as important for the nuclear power plant and the power lines connected to those substations and any other generating units that are considered to be important for the availability and reliability of electrical power for the nuclear power plant (e.g. generating units that may be used as alternative power sources for the nuclear power plant, as described in Section 2.3.6).

To guarantee that the development of the grid in the vicinity of the nuclear power plant will not have any adverse impact on the reliability of the connection between the nuclear power plant and the grid, it is a good practice to develop a legal framework prescribing that any changes that might happen on the grid elements that are identified as important for reliability in the vicinity of the nuclear power plant are known, assessed and agreed by the nuclear power plant operator. This legal framework may include an agreement, a methodology and a few procedures, as described in Section 2.4.3.

7. SPECIFIC DESIGN OF A NUCLEAR POWER PLANT TO WITHSTAND DEGRADED OFF-SITE POWER

Nuclear power plants rely on electrical power for various safety functions, and the reliability of the power supplies is important for the safety of the plant. Owing to the design of electrical power systems, all parts of the systems are normally connected regardless of their safety classification. The grid is part of the preferred power supply[16] for the nuclear power plant and the safety power systems. During power

---

[16] Preferred power is the power supply from the transmission system up to the safety classified electrical power system. It is composed of the transmission system, the switchyard, the main generator and the distribution system up to the safety classified electrical power system. Some portions of the preferred power supply are not part of the safety classification scheme.

operation, the power supply to the plant is normally provided through a generator, which dampens the variations arising from the grid.

The electrical power systems are necessary support systems for all levels of defence in depth. Any electrical event or disturbance that happens in the electrical power systems has to be handled in such a manner that the safety functions of the nuclear power plant can be fulfilled. The on-site power systems are linked together, and an electrical event on a non-safety bus will, in most cases, also affect the safety power systems.

The design bases for the on-site electrical power systems are fundamental for reliability and robustness. The design and operation of the on-site electrical power system need to anticipate that the power supply from the grid may become degraded or unavailable for many reasons, on either the grid side or the nuclear power plant side of the interface — for example, through failure of equipment, human error or short circuit faults (see Section 4.2). The design considerations for on-site electrical power systems need to cover all modes of operation and anticipated electrical events that could impact the reliability of the electrical power systems of the nuclear power plant, some of which are listed in SSG-34 [4], such as the following:

— Symmetrical and asymmetrical faults;
— Subsynchronous resonance phenomena;
— Large motor starts;
— Momentary perturbations in the grid system, such as switching surges or lightning strikes;
— Capacitor bank switching;
— Loss of transmission system elements, including single phase open conditions;
— Formation of grid islands and resulting frequency excursions and voltage excursions.

The design basis needs to account for the continuous operating ranges of voltage and frequency, all possible events that could cause transients (dynamic or continuous variations) and internal and external hazards that jeopardize the availability of the power supply to the plant.

7.1. PLANT DESIGN PROVISIONS FOR THE LOSS OF OFF-SITE POWER

Typically, the design features of a nuclear power plant are in place to cope with the anticipated unavailability of off-site power. Depending on the plant's electrical design and the plant–grid interface, there has to be a possibility to switch to another AC power source in case of a LOOP event.

Off-site power is normally supplied by at least two physically independent off-site circuits designed and located to minimize, to the extent practicable, the likelihood of their simultaneous failure. When the grid disturbances propagate to an event when power supply from the grid (i.e. main power line and backup power lines) and from the main generator become unavailable (i.e. the off-site power supplies and the main generator are lost), the necessary power to supply emergency switchgear buses is ensured by emergency diesel generators (EDGs), which start automatically at a signal indicating loss of voltage on the emergency switchgear bus.

7.1.1. Design for switching to house load operation

Some nuclear power plants are designed to operate in island mode so that the plant supplies power to its own auxiliary systems. This capability is available when the entire external load connected to the power plant is disconnected and there is no reactor or turbine trip. During this house load operating mode (see the Glossary), the reactor operates at a reduced power level (typically 5–10% of full power) that is sufficient to generate enough electrical power to supply the auxiliaries. This type of design affords an uninterrupted power supply for the house loads.

The house load operation capability is useful for performance and for safety aspects because:

— It provides an alternative electrical source different from the grid system and without the activation of on-site emergency power sources (e.g. EDGs).
— It keeps the reactor at power mode (i.e. without a hot or cold shutdown), so it can return to full power operation fairly quickly once the transmission system connection is restored.

To achieve house load operation, circuit breakers are necessary to separate the plant generator from the grid. This mode of operation, however, involves complex challenges for reactor systems due to reactivity feedback and control of the large step change required in reactor power. In addition, the load rejection results in a perturbation in the plant electrical system, which has to be designed to withstand the transient voltage and frequency variations (see Ref. [9]).

7.1.2. Fault clearing system and coordination of protection

To minimize the effects of any faults from the off-site (e.g. grid) and on-site electrical power systems, a coordination of protection and fault clearing systems is provided that will disconnect only the faulty equipment. Backup features are also provided for the event in which a primary protection feature or a fault clearing device fails. Since battery chargers, inverters and motor generator sets are generally sources of limited short circuit current, the coordination of the protective devices and the available fault current receives special attention. The coordination of protection is designed to work properly both during power operation and during shutdown conditions, as described in SSG-34 [4].

7.1.3. Power transfer capability

In most designs, when a fault occurs on unit transformers or when the unit generator trips, power to the nuclear power plant's auxiliary systems and equipment may be supplied from the grid system via standby transformers. This change in power supply requires fast transfer from the auxiliary transformer to the standby transformer and its grid connection. One of these connections is designed to be available within a few AC cycles following a loss of coolant accident to ensure that core cooling, containment integrity and other vital safety functions are maintained.

Fast transfer is required before low voltage or undervoltage is detected on the safety buses of the nuclear power plant. In the case of undervoltage, the safety buses and corresponding loads are disconnected from the remaining power grid, the diesel generators are initiated and safety loads are reconnected according to a predefined loading scheme. The available time for the transfer to the standby transformer is in the range of a few AC cycles (20–100 ms).

The transfer is normally automatic, but provisions are made to initiate the transfer also manually. Studies are performed to analyse the impact of the voltage, phase angle and frequency on buses and motors before, during and immediately after a bus is transferred. Reacceleration of motors also has to be considered in the study.

7.1.4. Design for loss of off-site power

When there is a simultaneous loss of preferred off-site AC power to all safety buses, this condition is called LOOP. (Note: it is likely that the non-safety buses will also lose power.) LOOP may result from events entirely within the nuclear power plant or from events external to the nuclear power plant.

The plant design basis considers the LOOP as an initiating event, which might occur either independently or as a concurrent event with a postulated initiating event. The expected frequency of the LOOP event is normally much greater than for other initiating events and transients that can challenge plant safety. Upon a LOOP event, the emergency switchgear buses are supplied from the on-site standby AC power sources such as EDGs.

Following a major interruption, the grid recovery by the grid system operator restores the power to essential services and to electricity consumers. This includes restoring off-site power supply to nuclear power plants as a priority, in a reasonably assured period. This period, which is typically assessed and determined by the grid system operator, is an essential consideration in the design of on-site AC power systems, as to the limitations and capabilities regarding grid restoration after the interruption of off-site power. For example, the mission time of the EDGs (including the fuel supply) is up to a month or so, with an expected recovery time and adequate margin.

7.1.5. Design for station blackout

The plant condition where there is complete loss of all AC power from the preferred off-site AC sources, from the main generator and from the standby AC power sources important to safety to the essential and nonessential switchgear buses is called an SBO. During an SBO, DC power supplies and uninterruptible AC power supplies may be available as long as batteries can supply the loads. Alternative AC power supplies need to be available.

Although the Fukushima Daiichi accident progressed well beyond the expected consequences of an SBO, many of the lessons learned from the accident remain valid. A failure of the plant power supply system such as the one that occurred at Fukushima Daiichi represents a design extension condition that requires management with predesigned contingency planning and operator training (see also Ref. [9]).

The coping time for an SBO[17] event for a specific nuclear power plant design determines the safety margin in the plant design when responding to the SBO event. The SBO coping time provides a timeline by which effective countermeasures have to be implemented to prevent fuel damage. In some nuclear power plant designs, the SBO coping time is very short, which complicates the implementation of effective measures to either restore the electrical power supply or activate an effective heat removal strategy.

A restoration of off-site power to the nuclear power plant is required much earlier when the on-site AC power supply sources (e.g. EDGs) cannot meet their anticipated and designed mission time and result in an SBO. Nuclear power plants are generally equipped with on-site alternative backup AC power sources (e.g. gas turbine generators or mobile diesel generators) and DC power sources (i.e. batteries and/or DC/AC inverters) to withstand an SBO for a limited period (the SBO coping time can vary between 4 and 72 h in different nuclear power plant designs) and support powering essential systems and components. The determination of the coping time, when this on-site equipment is used, is primarily based on the time that it would take to restore AC power sources to the nuclear power plant from the grid system and/or on the capacity and longevity of available measures for coping with an SBO.

Grid system faults resulting in a loss of AC power supply to the nuclear power plant and the recovery time for the grid supply are of great significance for nuclear safety, particularly for LOOP and SBO events.

The operators of the nuclear power plant have to enter into agreements with the grid system operator to ensure that appropriate priority[18] is given to restoration times for power supplies to a nuclear power plant to recover from power interruptions, particularly to recover from an SBO event within the coping time (i.e. when the preferred power source or standby AC power source is recovered).

7.1.6. Unanticipated degradation in the quality of off-site power

Operating experience has shown that the protections designed to cope with anticipated transients may sometimes fail to detect the degradation of power quality, and the disconnection of the nuclear power plant might not happen in an event that was not anticipated or considered in the design.

---

[17] Station blackout coping time is the time available from the loss of all AC power to the safety bus until the onset of core damage if no countermeasures are taken.

[18] From the grid system operator's point of view, it may not be the highest priority to restore supplies to a nuclear power unit that has been disconnected from the grid if it has tripped. See Ref. [6] for further details.

For example, an open phase condition[19] in an AC electrical system may result in consequences that have generally not been considered in the detection and protection schemes in the existing design of the electrical power systems of nuclear power plants (see Ref. [10] for further details).

An open phase condition may occur as a result of various faults, such as breaker poles failing to open or close, transformer bushings or line insulators breaking, and improperly connected conductors. This type of fault creates a voltage imbalance (and a current imbalance under load conditions) in the electrical power system. Operating experience has shown that open phase conditions may be difficult to detect under all operational conditions with existing plant instrumentation and electrical protection schemes, and there is a potential for a severe voltage imbalance which may result in the degradation or failure of electrical equipment. (For example, when a single or double open phase condition occurs on the high voltage side of large transformers under no load or lightly loaded conditions, it is likely that voltages will be regenerated, to some degree, in all phases on the high and low voltage side of the transformers.) In some cases, the inability to detect and disconnect the degraded power source from the safety bus has prevented the shift to a standby off-site power supply or standby AC power source.

To cope with degraded power, an assessment has to be performed on how to enhance independence between the power supplies to safety systems and the off-site grid. References [9, 14] provide further details on coping with degraded power. By design, a few provisions may bring more independence between off-site power sources. The provisions, which are mainly based on diversification and separation, can help to minimize the adverse impact of an unanticipated transient or to prevent its propagation to the power plant. Independence of the plant's electrical power systems can be achieved either by functional diversification or physical separation within the electrical system itself.

7.1.7. Diversity

Many plant designs are equipped with non-electric circulation or water injection systems (e.g. turbine driven or diesel driven feedwater pumps), which are independent of AC power supplies. However, these pumps may need a DC power supply to control speed and flow valves, or local manual operations may be required.

Diversified functions to power supplies can be achieved by powering the different trains of safety systems from different electrical sources during normal operation.

Electrical separation can also be provided by installing an electrical isolation device. For example, in some designs, control rod systems in pressurized water reactors or recirculation pumps in boiling water reactors are powered by motor–generator sets. In any of these cases, it needs to be proven that these provisions have the same or better reliability than an alternative distribution scheme of AC power from the electrical grid.

In addition to physical separation and electrical isolation, diversity might be necessary to increase independence between redundant systems or between systems supporting different levels of defence in depth. This may be achieved using diverse power sources or by providing supply from uninterruptible power supplies.

---

[19] An open phase condition is defined as an open circuit of one or two of the three phases of any power circuit.

8. MAINTAINING AND CONTINUOUSLY IMPROVING RELIABILITY AND RESILIENCE WITH EVOLVING FACTORS AND CONDITIONS

The reliability and resilience of nuclear power plants and their interconnection to the grid system may be affected by changes to internal or external factors. The internal factors could include changes to nuclear power plant operation arising from operational experience feedback, ageing and obsolescence of components, and changes in nuclear regulations. The external factors could include aspects that affect the electrical grid, such as changes in demand and changes in grid regulations. They can also include changes in the government's energy policies and changes in the nature of the electricity market.

The internal factors may be dependent on the external factors (e.g. changes in the governmental energy policy may affect nuclear energy development). Many external factors are linked with internal factors (e.g. the development of renewable energy may increase the need for the flexible operation of nuclear power plants). Changes may be gradual and continuous or abrupt. Whatever the origin of the changes, the possible consequences for the reliability and resilience of the nuclear power plants and their connections to the grid need to be assessed.

Nuclear regulations generally require the nuclear operator to periodically analyse the reliability and resilience of the nuclear power plant with the grid. This is the preferred framework for both nuclear and grid system operators to review together the observed or upcoming changes that may have consequences. The evolution of the nuclear power plant's reliability indicators detected in operation and due to changes in the electrical grid need to be addressed in the reviews of operating experience feedback.

The aim of this section is to raise awareness of those changes and evolutions of factors in the nuclear power domain, in the electrical grid domain and in terms of climate change that may have an adverse impact on the reliability and resilience of the electrical grid and its connections to nuclear power plants, so that Member States can evaluate their own situations and take any necessary actions.

8.1. CHANGES IN THE NUCLEAR ENERGY DOMAIN

This section introduces key developments in the nuclear energy domain that have implications for the operation of nuclear power plants and their interaction with the electrical grid. It focuses on the evolution of nuclear regulatory frameworks and on obsolescence and ageing management programmes, highlighting how continuous regulatory strengthening, periodic safety reviews and technological change shape policy decisions related to long term operation, safety enhancement and system resilience. These changes, while driven primarily by nuclear safety objectives, can also influence grid reliability, market structures and the strategic alignment between nuclear energy policy and broader power system development.

8.1.1. Evolution of nuclear regulation

The safety of a nuclear installation is not fixed, and the nuclear regulatory authority expects that improvement is constantly sought, taking account of experience feedback and new knowledge. The periodic reviews of existing facilities carried out by operators/licensees constitute a framework for verifying their compliance with existing and updated requirements, reassessing their level of safety, looking for possibilities for improvement and assessing the acceptability of their operation. The nuclear regulatory body may require certain modifications to the plant to allow continued operation.

The evolution of reference situations to assess nuclear power plant hazards is one of the domains where safety requirements have become more stringent.

Because the nuclear regulatory body is independent of the grid or electricity regulator, the changes in nuclear regulation are independent of the grid requirements, but may have beneficial effects on nuclear power plant reliability and the resilience of the grid.

8.1.2. Obsolescence and ageing programmes

There have been many changes since the first nuclear power plants were built. For instance, the technology of electrical grid equipment and digitalization have completely changed the design and operation of grids. Furthermore, the power generating units have increased size and integrated power electronics grid connection technologies. Additionally, large industrial electricity consumers connected directly to the transmission system are commonly used for grid balancing.

These changes have resulted in other changes. Using the same examples, larger generating units have fostered investments to interconnect electrical systems. Also, the information technology and digitalization benefits for grid design and operation have facilitated generation–transmission unbundling and the development of power markets. Additionally, better power quality has contributed greatly to the use of standardized electrical equipment and electronics (e.g. grid protections linked to plant transients).

8.2. CHANGES IN THE ELECTRICAL GRID DOMAIN

8.2.1. Grid operation and maintenance

8.2.1.1. Changing power flows

Operational planning by the grid system operator can cause changes in power flows at or near the substations where a nuclear power plant is connected. These changes in power flows during grid operation may be due to changes in the running arrangements in substations, reoptimization of power flows and preservation of operational margins of reactive power or stability. They may also be due to the integration of new transmission assets or due to grid outages for maintenance, or other factors. Such changes in power flows when the nuclear power plant is generating, or during nuclear power plant outages, may induce:

— Reduction of the short circuit level at the main or the standby connections below the minimum value requested in case of OLCs;
— Larger variations of reactive power and varying thermal gradients in the generator (leading to rotor vibrations).

All grid situations bringing changes in the power flows near substations where nuclear power plants are connected have to be detected and analysed, in close coordination and cooperation with the nuclear power plant operator and the grid system operator, to take action and to mitigate, if necessary, the consequences for the following:

— The reliability of the components of the nuclear power plant;
— The independence between normal and auxiliary external electrical sources;
— The reliability of the external electrical source;
— The operational availability of the off-site power;
— The measures to be taken in the event of loss of auxiliary sources;
— The measures to be taken in the event of LOOP.

The formal agreements between the grid system operator, the transmission system owner and the nuclear power plant operator need to ensure that they do not take any steps that could adversely affect nuclear safety, as discussed in Section 2.4.3.

In some Member States, the electrical grid was developed intensively 40–50 years ago and much of the original infrastructure is still in service. Assets are growing older and are more likely to become unreliable, and may jeopardize system reliability and security, while increasing the costs of operation and maintenance.

An important issue for a transmission system owner is to decide on a plan for the refurbishment and replacement of assets. A limited number of outages can be taken each year for asset replacement if the transmission system is to continue in normal operation during this work. Hence, there is a limit to the rate at which old equipment can be replaced, so a major refurbishment or replacement programme may have to be spread over many years.

Another significant challenge for the transmission system owner is changes in technology and in the supply chain, which make certain transmission assets and components obsolete. Spare or replacement parts for some equipment may cease to be available. This may force the transmission system owner to replace existing equipment before it reaches the end of its useful life.

Grid ageing and obsolescence issues need to be addressed during grid maintenance, which needs to be coordinated with nuclear power plant generation and outage planning according to the agreements between the nuclear power plant operator, the transmission system owner and the grid system operator, as discussed in Section 2.4.3.

Replacement of equipment at or close to the substation to which the nuclear power plant is connected needs to be done in consultation with the nuclear power plant operator. The way in which the replacement is to be carried out needs to be subject to agreement with the nuclear power plant operator. This is to ensure that the new equipment, or the outages required for its installation, do not adversely affect the reliability of the grid connections, or otherwise affect the nuclear licensing basis.[20] The formal agreements between the grid system operator, the transmission system owner and the nuclear power plant operator need to allow for such steps, as discussed in Section 2.4.3.

8.2.2. Changes in power generation

Owing to developments in technology, changes in market conditions, stronger environmental considerations and associated energy policies, the generation mix is changing in many Member States. These changes in the mix have recently included shifting from conventional thermal power plants to variable and intermittent renewable energy sources and retiring coal fired generating units and replacing some with natural gas fired generating units. Changes to the operational characteristics of generating units affect system planning and frequency stability.

8.2.2.1. Generation versus demand balancing

The system frequency varies because of stochastic and deterministic variations in power generation and demand, and this variation needs to be compensated by dispatchable generating units. The inherent variability of the output of renewable energy sources, such as wind and solar power, also adds to the variability of system frequency. In particular, a nuclear power plant may be requested to vary its output over a power range as frequently as the limits granted in the OLCs allow. The nuclear power plant output may also be reduced by grid situations with excess power generation from non-dispatchable units. In a system with an electricity market, these situations can lead to negative electricity prices at certain times and potential stability problems. Operating experience from nuclear power plants has shown that frequent load following decreases the operational availability compared with baseload operation, although the

---

[20] For example, in France, the nuclear regulator has asked to requalify and to monitor the gas insulated components of the substations to which nuclear power plants are connected during the ten-cycle inspection period.

effect is not large. Clear regulations may need to be defined in some countries to give the grid system operator the right to control electricity generation that is currently non-dispatchable.

Renewable energy sources connected to distribution networks may provide less reactive power support in the transmission system[21] as conventional power generation is decommissioned. This is of relevance for high photovoltaic generation and moderate load conditions. This results in longer voltage recovery times after faults and a higher risk of voltage collapse. High voltage situations (i.e. a voltage greater than the normal operation limit, usually defined as 1.05 times the nominal voltage) may happen at the connection points, degrading the reliability of electrical equipment and reducing its lifetime. Continuous monitoring of the voltage at connection points with alerts sent to the grid system operator are necessary to limit these situations.

8.2.2.2. System inertia

System inertia is mainly provided by the rotors of the operating synchronous machines, such as turbogenerators and hydroelectric generators. Some inertia is also provided by large induction motors in the consumer demand. Different power generating technologies provide different amounts of inertia, so that a change in the generation mix in a system can change the system inertia and hence change the behaviour of the system frequency after the trip of a large power generating unit.

Changes in generation sources can lead to a reduction in system inertia for the following reasons:

— Typically, open cycle gas turbine generators and gas fired reciprocating engines have a lower inertia constant than existing, synchronous generating units.
— Renewable power generation technologies (e.g. large wind turbines, solar photovoltaic systems) of standard design do not directly contribute any inertia to the system.
— The inverters of HVDC connections (effectively acting as generation sources) also do not contribute any inertia.

By contrast, the large, four pole synchronous generating units used in large modern nuclear power units generally have a higher inertia constant than the generating units in large coal or oil fired power plants.

A reduction in system inertia means that it will be harder to control the system frequency after the sudden loss of large generating units, such as nuclear power units, resulting in abnormally low frequencies, which is discussed in detail in Ref. [6]. In several Member States, the protection settings on small power generating units have been changed to reduce the risk of their disconnection during a low frequency transient. These include the settings for ROCOF relays and low frequency trip relays.

The grid system operator will need to be aware of this potential reduction in system inertia and take measures to provide a more rapid control of the frequency. It may also be possible for some designs of wind turbines or HVDC inverter systems to provide simulated inertia by a modification to their control systems. A recent technological development is the use of fast response battery inverter systems to provide simulated inertia.

8.2.3. Changes in demand

In many Member States, consumer electricity demand is increasing, and such increases are likely to continue for many years. To maintain a reliable system with continually increasing demand, it is necessary to constantly add generation capacity and make additions and reinforcements to the transmission system to ensure that the demand can be met with suitable means to avoid overload.

In addition to changes in the magnitude of demand, the character of the demand may also change. This could include changes in the relative demand between winter and summer or changes in the

---

[21] Small generation facilities will not influence system operation individually, but their aggregate impact may be significant.

demand profile during the day. This may be caused partly by changes to the climate, as also mentioned in Section 8.3. The grid system operator will need to monitor such changes and adjust system planning to accommodate them. Some Member States have also observed a gradual change in the power factor of consumer load, becoming less inductive and more capacitive, so that it tends to increase voltages on the transmission system. This requires a change in voltage control policies to ensure that the transmission system voltage remains within the required range.

There has been increasing research (see Refs [83–88]) that shows unidirectional, bidirectional or multidirectional relationships between all segments of an electrical grid (e.g. the choice of generation type, extent and structure of transmission system, electricity consumption), economic growth, population growth and/or a demographic shift, and urbanization, as well as changes in the type of economy (e.g. change from manufacturing to service economy and vice versa) and changes in high demand users, which can also support grid reliability from the demand management side, such as the paper industry, the aluminium and steel industries, and mining installations, and their spatial distribution. Although these studies show different and mixed impacts on electrical grid planning, design and operation for both developed and developing countries, such impacts need to be considered in terms of the potential adverse or benign effects on grid security, reliability and resilience on a case by case basis.

8.2.4. Changes in energy policy and regulations

Historically, any change to the regulation of the system for electricity services (i.e. generation, transmission and distribution) has changed the behaviour and functioning of system participants, including nuclear power plants. Such changes, in turn, have had various direct or indirect impacts (benign or adverse) on the physical and economic conditions of the electrical grid (see Ref. [89]). These impacts, primarily on system design, operation and maintenance, as well as on electricity generation and demand, affect the management of the reliability and resilience of the electrical grid.

The regulatory scope, requirements, framework, implementation and stakeholders have been evolving (particularly since the 1990s) and are expected to continue to evolve. Therefore, the effects of regulatory changes on the nuclear power plant's interface with the grid and the maintenance of the grid's reliability and resilience will vary and will need to be reassessed, readjusted or refined.

The following sections discuss two examples of impacts on the management of reliability and resilience of the electrical grid for potential future changes in the regulations.

8.2.4.1. Ownership, responsibilities and investment on infrastructure

In many Member States, the electrical grid and the generating stations were developed under a utility entity that was under either public (e.g. government, cooperative, municipality) or private (e.g. investor, entrepreneur) ownership. Regardless of the ownership, these utility companies were often vertically integrated, with both transmission and generation (and distribution) functions belonging to the same utility. The reliability reserves were arranged among neighbouring utilities that had a similar structure. Typically, the publicly owned utilities had little or no regulation (as the owner, voter and consumer were essentially the same and worked towards their own interest) in setting the infrastructure and operating it to ensure that the electricity was available reliably with the best retail rates. On the other hand, privately owned utilities were regulated in setting the rates to be charged for their services and ensuring proper planning, design and structuring of the system and its operation and maintenance (which were considered to be capital costs) and operating expenses (which were considered in the regulated rate). Hence, the power plants and the transmission system could be planned, designed and operated in a coordinated way.

Since approximately 1990, various factors have resulted in high electricity rates and consequent regulatory expansion, which were not working well, and an increasing number of Member States have made significant changes to their regulatory structures (commonly referred to as 'deregulation'), typically in the following ways:

— State owned utilities were privatized and/or competitive market forces were imposed on privately owned utilities.
— Private companies were separated horizontally for electricity services (i.e. generation, transmission and distribution).
— New institutions were established for competitive markets (or to mimic those; e.g. power pools in the United Kingdom) and for the horizontal separation of services (e.g. independent operators of the electrical grid, such as the independent system operators in the USA).

The long term commitment to deliver energy to consumers by the utilities brings more focus on the system's security of supply to the consumer. Owing to this factor, in a regulated system, it is ensured (and is easier) that vertically integrated (both public and private) utilities build, operate and maintain a reliable electrical grid in the service area from the following two perspectives:

(a) The grid system operator and transmission system owner can work together with the generating units (both owned by the same utility in principle) in the stabilization and reliability of the electrical grid using the inherent characteristic of the transmission system and generating units, as well as in the planning, design and optimization of an integrated generation, transmission and distribution infrastructure.
(b) The cost of infrastructure improvements for system security, reliability and resilience can be compensated by the consumer (either by their vote in cases of the publicly owned utilities or by regulatory oversight in cases of privately owned ones).

On the other hand, investment in infrastructure may be more complicated in a deregulated energy service scheme because of the ownerships, responsibilities and capacities of different services. For example:

— New power stations and other generation sources will be built by the power generation company's owners/investors.[22]
— Transmission system improvements are approved by the transmission system owner.[23]
— The grid system operator and transmission system owner are responsible for improvements to the overall operation and maintenance of the electrical grid for security, reliability and resilience.

All these ownerships, responsibilities and investments for needs would have to be aligned and agreed upon by all stakeholders, the decision for which would be driven by the electricity and service market conditions and by the recovery of costs or other financial incentives that directly affect the security, reliability and resilience of the electrical grid. For example, new investment in the transmission system may be justified by financial reasons that affect all stakeholders, such as a reduction in transmission losses or a reduction in constraints (e.g. the available transmission capacity) that prevent the operation of generating units and challenge control of the system. However, for all other more likely cases where there is no common financial justification, it would be increasingly desirable that the grid system operator and transmission system owner have rules and standards that require the additions and reinforcements to the generating units and transmission system to continue to meet, or improve on, existing standards for the security, reliability and resilience of services.

---

[22] Similarly, the owners of existing power generators may decide to close permanently if electricity and services market conditions are not favourable.

[23] Some of them own newer systems — therefore, the cost of system improvements will be higher than for the old systems, which had lower initial costs and a longer depreciation period.

8.2.4.2. Competitive market based prices and incentives for electricity and services

Electricity price is regulated to prevent privately owned utilities (which are monopolistic in providing electricity services) from overcharging consumers for a higher profit, particularly at times of high demand and low electricity supply. This risk would exist in competitive markets and in vertically integrated or horizontally separated private utilities (e.g. in the case of the owners of the transmission systems, who typically have a natural monopoly and market power through their control of access to the transmission lines).

Historically, electricity price was cost based, and wholesale and retail prices were set/overseen by the regulators. Further, in most regulated electricity markets, there was no payment to power plants for providing ancillary services (e.g. for the control of voltage and frequency) that are essential for reliable control of the electrical grid.

Until the oil embargo in the 1970s, increased electricity demand led the utilities to build more and larger generating units, which continued afterwards, although with the addition of large generating units with either low fuel costs, such as nuclear power plants, or domestic fuel supply, such as coal. However, later, in the early 1980s, due to declining demand and overbuilt supply, most utilities often needed rate increases to recover the costs of already built supply and to avoid bankruptcy. This led to the restructuring and deregulation of electricity markets.

The 'market based' and 'cost of service' based price of electricity determined the power exchange in the 'energy only' markets. Power exchanges were, then or later, started in several related markets, such as 'day ahead', 'hour ahead' and 'real time' markets.

To ensure sufficient integrity, stability and reliability (not only the generation capability) of the electrical power system, other 'important to grid system operation' aspects of generating units, such as readiness for power generation, support for grid stability and reliability services, started to be valued — more so in response to the rapid change of energy mix to include intermittent renewable generating units and the need for more flexibility. This led many Member States to start trading these ancillary services using 'capacity markets' of some kind and/or establishing requirements in grid codes for the minimum capability of generating units to provide those services (see Ref. [90]).

In highly deregulated markets, it has become inevitable for the grid system operator to establish and impose the requirements, penalties or rewards on the generating units to support a secure and reliable system operation, as there may be no obligations and/or incentives for generating units to support the system's security of supply and it is up to the market to give signals for investments.

More recently, in order to meet the energy policy aspects of climate change targets, carbon emission caps, tariffs or credits were established in some Member States, resulting in the operation of secondary markets for exchanging carbon emissions and so forth.

These changes in market regulations have had a direct impact on nuclear power plants, which were mostly built during a regulated market but are now operating in deregulated markets that are much more volatile and diverse, and thus react promptly to price signals. In addition to being low cost fuel, baseload generating units, nuclear power plants have become more visible and valued for other benefits, such as the capability to perform load following and frequency control, and more importantly, generating large amounts of low carbon emission electricity. As learned from the past, the energy markets will continue to change, affecting the nuclear power plant generation environment, but also affecting the operation strategy, which needs to be adjusted and refined in order to support (or rely on) grid reliability and resilience.

8.3. CHANGES IN CLIMATE

It is the scientific consensus that there will be noticeable changes to the world's climate during the twenty-first century, which will lead to changes in global and regional mean and extreme weather conditions and to the number and severity of extreme weather events. These changes may be different for different geographical regions (see Refs [91–93]). The impact of these changes on the energy sector will vary depending on the sources of and the technologies used for power generation, the transmission infrastructure and demand patterns. These are discussed in detail in Ref. [94]. For example, gradual climate change may lead to a change in:

— The relative frequency of faults on transmission systems due to the changing number and severity of weather events (e.g. temperature, precipitation, wind, thunderstorms, with increases in some areas and reductions in others);
— The availability and efficiency of generating units;
— The patterns of electricity demand for heating and cooling due to changes in the number of extremely warm or cold days.

Faults on the transmission system caused by weather events are discussed in Section 4.2.1. Changing weather events can also affect the reliability of the grid system by affecting the operation of power plants. The measures needed to mitigate the effect of climate change may also be different in different geographical areas. An assessment in the United Kingdom (see Refs [54, 55]) concluded that there is no immediate need for an increase in standards relating to wind, but that a significant number of substations will need improved flood defences.

The impacts on reliability can be direct (i.e. impact on a given facility, a type of facility or an infrastructure component) and indirect (i.e. affecting part(s) of the system or the environment that impact the overall system or other parts of the system). For example, Ref. [94] indicates that:

— Icy rain can damage solar panels, or an ice storm may damage transmission towers and lines, or extreme winds could destroy wind turbines, as a direct effect. Damaged transmission lines, in turn, may force an otherwise intact nuclear power plant to shut down because it cannot export electricity to the grid system — an indirect effect.
— A drought leading to low water levels in a river can directly impact the cooling of a nuclear power plant, forcing it to reduce or cease power generation. It also may indirectly lead to the shutting down of a coal power plant because coal delivery via river is disrupted.

The effects of such direct and indirect impacts on the reliability of the electrical power system as a whole and on the reliability of connections to nuclear power plants will need to be assessed to ensure the continuing safety of the nuclear power plant. This may include a reassessment of the national or international standards that are used in the design of equipment for nuclear power plants and transmission networks.

To assist in such an assessment, Ref. [94] presents a detailed discussion of the many possible direct and indirect consequences of climate change.

REFERENCES

[1]   INTERNATIONAL ATOMIC ENERGY AGENCY, Interaction of Grid Characteristics with Design and Performance of Nuclear Power Plants, Technical Reports Series No. 224, IAEA, Vienna (1983).

[2]   INTERNATIONAL ATOMIC ENERGY AGENCY, Introducing Nuclear Power Plants into Electrical Power Systems of Limited Capacity, Technical Reports Series No. 271, IAEA, Vienna (1987).

[3]   INTERNATIONAL ATOMIC ENERGY AGENCY, Safety of Nuclear Power Plants: Design, IAEA Safety Standards Series No. SSR-2/1 (Rev. 1), IAEA, Vienna (2016).

[4]   INTERNATIONAL ATOMIC ENERGY AGENCY, Design of Electrical Power Systems for Nuclear Power Plants, IAEA Safety Standards Series No. SSG-34, IAEA, Vienna (2016).

[5] INTERNATIONAL ATOMIC ENERGY AGENCY, OECD NUCLEAR ENERGY AGENCY, Nuclear Power Plant Operating Experiences from the IAEA/NEA Incident Reporting System 2002–2005, NEA No. 6150, OECD, Paris (2006).

[6] INTERNATIONAL ATOMIC ENERGY AGENCY, Electrical Grid Reliability and Interface with Nuclear Power Plants, IAEA Nuclear Energy Series No. NG-T-3.8, IAEA, Vienna (2012).

[7] EUROPEAN COMMISSION JOINT RESEARCH CENTRE, Summary Report on Events Related to Loss of Offsite Power and Station Blackout at NPPs, JRC100227, Publications Office of the European Union, Luxembourg (2016).

[8] DEPARTMENT OF ENERGY, Analysis of Loss-of-Off-site-Power Events 1987–2015, Report No. INL/EXT-16-39575, Idaho National Laboratory, Idaho Falls, ID (2016).

[9] INTERNATIONAL ATOMIC ENERGY AGENCY, Design Provisions for Withstanding Station Blackout at Nuclear Power Plants, IAEA-TECDOC-1770, IAEA, Vienna (2015).

[10] INTERNATIONAL ATOMIC ENERGY AGENCY, Impact of Open Phase Conditions on Electrical Power Systems of Nuclear Power Plants, Safety Reports Series No. 91, IAEA, Vienna (2016).

[11] INTERNATIONAL ATOMIC ENERGY AGENCY, Nuclear Power Plant Operating Experience from the IAEA/NEA International Reporting System for Operating Experience (IRS), EDG Failed to Start after Undetected Loss of Two Phases on 400 kV Incoming Offsite Supply, IRS No. 8315, IAEA, Vienna (2015).

[12] NUCLEAR REGULATORY COMMISSION, Design Vulnerability in Electric Power System, NRC Bulletin 2012-01, Office of New Reactors, Washington, DC (2012).

[13] INTERNATIONAL ATOMIC ENERGY AGENCY, OECD NUCLEAR ENERGY AGENCY, International Reporting System for Operating Experience (IRS), Loss of 400 kV and Subsequent Failure to Start Emergency Diesel Generators in Sub A and Sub B, IRS No. 7788, IAEA, Vienna (2006).

[14] LUNDBÄCK, M., KARLSSON, M., Degraded Power Supplies (Proc. 24th Int. Conf. Nucl. Eng. (ICONE24), Charlotte, 2016), American Society of Mechanical Engineers, New York (2016).

[15] INTERNATIONAL ATOMIC ENERGY AGENCY, Operational Limits and Conditions and Operating Procedures for Nuclear Power Plants, IAEA Safety Standards Series No. NS-G-2.2, IAEA, Vienna (2000).

[16] INTERNATIONAL ATOMIC ENERGY AGENCY, The Fukushima Daiichi Accident, IAEA, Vienna (2015).

[17] Strengthening the Agency's activities related to nuclear science, technology and applications, GC(67)/RES/10, IAEA, Vienna (2023),
https://www.iaea.org/sites/default/files/gc/gc67-res10.pdf

[18] INTERNATIONAL ATOMIC ENERGY AGENCY, Safety Classification of Structures, Systems and Components in Nuclear Power Plants, IAEA Safety Standards Series No. SSG-30, IAEA, Vienna (2014).

[19] INTERNATIONAL COUNCIL ON LARGE ELECTRIC SYSTEMS, The Future of Reliability — Definition of Reliability in Light of New Developments in Various Devices and Services Which Offer Customers and System Operators New Levels of Flexibility, CIGRE Technical Brochure 715, CIGRE, Paris (2018).

[20] SKARVELIS-KAZAKOS, S., et al., Resilience of Interdependent Critical Infrastructure, Working Group Report No. WGR_320_1, CIGRE, Paris (2021).

[21] INTERNATIONAL ATOMIC ENERGY AGENCY, Safety of Nuclear Power Plants: Commissioning and Operation, IAEA Safety Standards Series No. SSR-2/2 (Rev. 1), IAEA, Vienna (2016).

[22] INTERNATIONAL ATOMIC ENERGY AGENCY, Initiating Nuclear Power Programmes: Responsibilities and Capabilities of Owners and Operators, IAEA Nuclear Energy Series No. NG-T-3.1 (Rev. 1), IAEA, Vienna (2020).

[23] EUROPEAN ATOMIC ENERGY COMMUNITY, FOOD AND AGRICULTURE ORGANIZATION OF THE UNITED NATIONS, INTERNATIONAL ATOMIC ENERGY AGENCY, INTERNATIONAL LABOUR ORGANIZATION, INTERNATIONAL MARITIME ORGANIZATION, OECD NUCLEAR ENERGY AGENCY, PAN AMERICAN HEALTH ORGANIZATION, UNITED NATIONS ENVIRONMENT PROGRAMME, WORLD HEALTH ORGANIZATION, Fundamental Safety Principles, IAEA Safety Standards Series No. SF-1, IAEA, Vienna (2006),
https://doi.org/10.61092/iaea.hmxn-vw0a

[24] INTERNATIONAL ATOMIC ENERGY AGENCY, IAEA Nuclear Safety and Security Glossary, IAEA, Vienna (2022),
https://doi.org/10.61092/iaea.rrxi-t56z

[25] INTERNATIONAL ATOMIC ENERGY AGENCY, Building a National Position for a New Nuclear Power Programme, IAEA Nuclear Energy Series No. NG-T-3.14, IAEA, Vienna (2016).

[26] INTERNATIONAL ATOMIC ENERGY AGENCY, Non-Baseload Operation in Nuclear Power Plants: Load Following and Frequency Control Modes of Flexible Operation, IAEA Nuclear Energy Series No. NP-T-3.23, IAEA, Vienna (2018).

[27] EUROPEAN UTILITY REQUIREMENTS ORGANISATION, European Utility Requirements for LWR Nuclear Power Plants, European Utilities Requirement Document, Rev. D, EUR Organisation, Lyon (2012).

[28] ELECTRIC POWER RESEARCH INSTITUTE, Advanced Nuclear Technology: Advanced Light Water Reactor Utility Requirements Document (Rev. 13), 3002003129, EPRI, Palo Alto, CA (2014).

[29] NUCLEAR REGULATORY COMMISSION, Operating Experience Assessment — Effects of Grid Events on Nuclear Power Plant Performance, NUREG-1784, Office of Nuclear Regulatory Research, Washington, DC (2003).

[30] EUROPEAN NETWORK OF TRANSMISSION SYSTEM OPERATORS FOR ELECTRICITY, Guidelines for the Classification of Grid Disturbances Above 100 kV, ENTSO-E, Brussels (2017).

[31] INTERNATIONAL ELECTROTECHNICAL COMMISSION, International Electrotechnical Vocabulary (IEV) — Part 192: Dependability, IEC Standard 60050, IEC, Geneva (2015).

[32] INSTITUTE OF ELECTRICAL AND ELECTRONICS ENGINEERS, IEEE Standard Terms for Reporting and Analyzing Outage Occurrences and Outage States of Electrical Transmission Facilities, IEEE Standard 859-2018, IEEE, New York (2019).

[33] NORTH AMERICAN ELECTRIC RELIABILITY CORPORATION, Glossary of Terms Used in NERC Reliability Standards, NERC, Atlanta, GA (2023).

[34] WARD, D.M., The effect of weather on grid systems and the reliability of electricity supply, Clim. Change **121** (2013) 103–113,
https://doi.org/10.1007/s10584-013-0916-z

[35] HINES, P., APT, J., TALUKDAR, S., Large blackouts in North America: Historical trends and policy implications, Energy Policy **37** (2009) 5249–5259,
https://doi.org/10.1016/j.enpol.2009.07.049

[36] DAVIDSON, R.A., LIU, H., SARPONG, I.K., SPARKS, P., ROSOWSKY, D.V., Electric power distribution system performance in Carolina hurricanes, Nat. Hazards Rev. **4** (2003) 36–45,
https://doi.org/10.1061/(ASCE)1527-6988(2003)4:1(36)

[37] WINKLER, J., DUEÑAS-OSORIO, L., STEIN, R., SUBRAMANIAN, D., Performance assessment of topologically diverse power systems subjected to hurricane events, Reliab. Eng. Syst. Saf. **95** (2010) 323–336,
https://doi.org/10.1016/j.ress.2009.11.002

[38] REED, D.A., Electric utility distribution analysis for extreme winds, J. Wind Eng. Ind. Aerodyn. **96** (2008) 123–140,
https://doi.org/10.1016/j.jweia.2007.04.002

[39] VOLKANOVSKI, A., Wind generation impact on electricity generation adequacy and nuclear safety, Reliab. Eng. Syst. Saf. **158** (2017) 85–92,
https://doi.org/10.1016/j.ress.2016.10.003

[40] BRITISH STANDARDS INSTITUTE, Overhead Electrical Lines Exceeding AC 1 kV — General Requirements — Common Specifications, BS EN 50341-1:2012, BSI, London (2013).

[41] BAYLISS, C.R., HARDY, B.J., Transmission and Distribution Electrical Engineering, 4th edn, Elsevier, Amsterdam and London (2011).

[42] ELECTRIC POWER RESEARCH INSTITUTE, Overhead Transmission Line Lightning and Grounding Reference Book 2023 (Gray Book), 3002026976, EPRI, Palo Alto, CA (2023).

[43] ELECTRIC POWER RESEARCH INSTITUTE, T&D System Design and Construction for Enhanced Reliability and Power Quality, 1010192, EPRI, Palo Alto, CA (2006).

[44] INSTITUTE OF ELECTRICAL AND ELECTRONICS ENGINEERS, IEEE Guide for Improving the Lightning Performance of Transmission Lines, IEEE Standard 1243-1997, IEEE, New York (1997).

[45] ELECTRIC POWER RESEARCH INSTITUTE, AC Flashovers on Henan Power 500 kV Lines During Rain, 1013244, EPRI, Palo Alto, CA (2007).

[46] BROWN, R.E., Electric Power Distribution Reliability, 2nd edn, CRC Press, Boca Raton, FL (2008).

[47] AQUINO, J., DO PRADO, J.C., NAZARIPOUYA, H., BERTOLETTI, A.Z., Enhancing power grid resilience against ice storms: State-of-the-art, challenges, needs, and opportunities, IEEE Access **11** (2023) 60792–60806,
https://doi.org/10.1109/ACCESS.2023.3286532

[48] ELECTRIC POWER RESEARCH INSTITUTE, Distribution Grid Resiliency: Storm Response Practices, 3002006784, EPRI, Palo Alto, CA (2015).

[49] NORTH AMERICAN ELECTRIC RELIABILITY CORPORATION, 1998 System Disturbances — Review of Selected Electric System Disturbances in North America, NERC, Princeton, NJ (2001).

[50] BROSTRÖM, E., SÖDER, L., Ice Storm Impact on Power System Reliability, (Proc. 12th Int. Workshop Atmos. Icing Struct. (IWAIS XII), Yokohama, 2007), (2007).

[51] MARKOSEK, J., Severe Freezing Rain in Slovenia, Newsletter of the WGCEF No. 20, The European Forecaster, Météo-France, Trappes (2015).

[52] CENTRAL ELECTRICITY AUTHORITY, Report of the Inquiry Committee on Grid Incident in Northern Region on 7th and 9th March, 2008, Northern Regional Power Committee Katwaria Sarai, New Delhi (2008).

[53] ESKOM/AFRICAN CENTRE FOR ENERGY AND ENVIRONMENT, The Management of Wildlife Interactions with Overhead Power Lines, Southern Africa Power Pool Environmental Subcommittee Training Manual, Eskom, Johannesburg (2003).

[54] NATIONAL GRID ELECTRICITY TRANSMISSION, Climate Change Adaptation Report, NGET, Warwick (2021).

[55] ENERGY NETWORKS ASSOCIATION, 3rd Round Climate Change Adaptation Report, ENA, London (2021).

[56] HAYNES, W.M., CRC Handbook of Chemistry and Physics, 91st edn, CRC Press, Boca Raton, FL (2010).

[57] SCHIFF, A., Guide to Improved Earthquake Performance of Electric Power Systems, National Institute of Standards and Technology Report NIST GCR 98-757, Gaithersburg, MD (1998),
https://nehrpsearch.nist.gov/static/files/NIST/PB98177892.pdf

[58] SCHIFF, A., Guide to Improved Earthquake Performance of Electric Power Systems, ASCE Manuals and Reports on Engineering Practice No. 96, American Society of Civil Engineers (ASCE), Reston, VA (1999).

[59] INSTITUTE OF ELECTRICAL AND ELECTRONICS ENGINEERS, IEEE Recommended Practice for Seismic Design of Substations, IEEE Standard 693-2018, IEEE, New York (2019).

[60] DEPARTMENT OF HOMELAND SECURITY, Geomagnetic Storms — An Evaluation of Risks and Risk Assessment, Office of Risk Management and Analysis, Washington, DC (2011).

[61] ORGANISATION FOR ECONOMIC CO-OPERATION AND DEVELOPMENT, Geomagnetic Storms: OECD/IFP Futures Project on Future Global Shocks, OECD, Paris (2011).

[62] E&E NEWS, Electromagnetic Pulse: Effects on the US Power Grid, Executive Summary, E&E News Legacy Assets, Washington, DC (2011),
https://legacy-assets.eenews.net/open_files/assets/2011/08/26/document_gw_02.pdf

[63] DEPARTMENT OF HOMELAND SECURITY, Protecting the Electrical Grid from the Potential Threats of Solar Storms and Electromagnetic Pulse, Hearing before the Committee on Homeland Security and Governmental Affairs, US Government Publishing Office, Washington, DC (2016).

[64] NORTH AMERICAN ELECTRICRIC RELIABILITY CORPORATION, Transmission System Planned Performance for Geomagnetic Disturbance Events, TPL-007-3, NERC, Atlanta, GA (2019).

[65] PARLIAMENT OF THE UNITED KINGDOM, "Display of licence", Scrap Metal Dealers Act 2013, 2013, UK Government, London (2013), Section 10.

[66] NORTH AMERICAN ELECTRIC RELIABILITY CORPORATION, Cyber Security — Bulk Electricity System (BES) Cyber System Categorisation, CIP-002-5.1a, NERC, Atlanta, GA (2016).

[67] NORTH AMERICAN ELECTRIC RELIABILITY CORPORATION, Cyber Security — Electronic Security Perimeter(s), CIP-005-5, NERC, Atlanta, GA (2016).

[68] NORTH AMERICAN ELECTRIC RELIABILITY CORPORATION, Cyber Security — Physical Security of Bulk Electricity System (BES) Cyber Systems, CIP-006-6, NERC, Atlanta, GA (2016).

[69] INTERNATIONAL ATOMIC ENERGY AGENCY, Computer Security at Nuclear Facilities, IAEA Nuclear Security Series No. 17, IAEA, Vienna (2011).

[70] INTERNATIONAL ATOMIC ENERGY AGENCY, Computer Security for Nuclear Security, IAEA Nuclear Security Series No. 42-G, IAEA, Vienna (2021).

[71] INTERNATIONAL ATOMIC ENERGY AGENCY, Computer Security Techniques for Nuclear Facilities, IAEA Nuclear Security Series No. 17-T (Rev. 1), IAEA, Vienna (2021).

[72] INTERNATIONAL ATOMIC ENERGY AGENCY, Assessment of Defence in Depth for Nuclear Power Plants, Safety Reports Series No. 46 (Rev.1), IAEA, Vienna (2024),
https://doi.org/10.61092/iaea.dbwn-89a9

[73] EUROPEAN COMMISSION, Establishing a Network Code on Electricity Emergency and Restoration, Commission Regulation 2017/2196, EC, Brussels (2017).

[74] INTERNATIONAL ATOMIC ENERGY AGENCY, Format and Content of the Safety Analysis Report for Nuclear Power Plants, IAEA Safety Standards Series No. SSG-61, IAEA, Vienna (2021).

[75] INTERNATIONAL COUNCIL ON LARGE ELECTRIC SYSTEMS, Definition and Classification of Power System Stability, ELT_208_10, CIGRE, Paris (2003).

[76] KUNDUR, P.S., MALIK, O.P., Power System Stability and Control, 2nd edn, McGraw-Hill, New York (2022).

[77] McDONALD, J., Electric Power Substations Engineering, 3rd edn, CRC Press, Boca Raton, FL (2012).

[78] KUROI, K., OSHIDA, H., KOMATSU, C., KAWASAKI, Y. TOI, M., NAKATANI, H., Applications and Developments for the Remote Access to Protection Relays, CIGRE Session Materials B5-210-2012, CIGRE, Paris (2012).

[79] INTERNATIONAL COUNCIL ON LARGE ELECTRIC SYSTEMS, Air Insulated Substation Design for Severe Climate Conditions, CIGRE Technical Brochure 614, CIGRE, Paris (2015).

[80] INTERNATIONAL ELECTROTECHNICAL COMMISSION, Communication Networks and Systems for Power Utility Automation — Part 3: General Requirements, IEC 61850-3, IEC, Geneva (2013).

[81] ARZUL, J.Y., et al., Le Système Nerveux du Réseau Français Transport d'Electricité: 1946–2006: 60 Années de Contrôle Électrique, Tec & Doc Lavoisier, Paris (2012).

[82] WORLD ASSOCIATION OF NUCLEAR OPERATORS, Loss of Grid — Addendum, WANO Significant Operating Experience Report (SOER) 1999-1, WANO, London (2004).

[83] ZHOU, L., MAHALIK, M.K., PADHAN, H., GUPTA, M., GOZGOR, G., Effects of Age Dependency and Urbanization on Energy Demand in BRICS: Evidence from the Machine Learning Estimator, Article 749065, Frontiers in Energy Research, Lausanne (2021).

[84] WANG, N., FU, X., WANG, S., Economic growth, electricity consumption, and urbanization in China: A tri-variate investigation using panel data modeling from a regional disparity perspective, J. Clean. Prod. **318** (2021) 128529,
https://doi.org/10.1016/j.jclepro.2021.128529

[85] MINH, P.K., FICHTNER, W., McKENNA, R., The effects of urbanization on energy consumption and greenhouse gas emissions in Asean countries: A decomposition analysis, (Proc. 15th Int. Assoc. Energy Econ. (IAEE) European Conference, Vienna, 2017), IAEE, Vienna (2017).

[86] LIU, S., LI, H., Does financial development increase urban electricity consumption? Evidence from spatial and heterogeneity analysis, Sustainability **12** (2020) 1–17,
https://doi.org/10.3390/su12177011

[87] HONG, H., ZHENG, P., ZHU, L., ZHAO, Y., Research on impact mechanism of demand side of urban residents' electricity consumption: Analysis based on microscopic survey data, Complexity **2021** (2021) 1–15,
https://doi.org/10.1155/2021/5552516

[88] NATHANIEL, S.P., BEKUN, F.V., Electricity consumption, urbanization and economic growth in Nigeria: New insights from combined cointegration amidst structural breaks, Research Africa Network (RAN) Working Paper 20/013, J. Public Aff. **21** (2021) e2102,
https://doi.org/10.1002/pa.2102

[89] DEPARTMENT OF ENERGY, A Primer on Electric Utilities, Deregulation, and Restructuring of US Electricity Markets, PNNL-13906, Office of Energy Efficiency and Renewable Energy, Washington, DC (2002).

[90] INTERNATIONAL ENERGY AGENCY, Re-powering Markets — Market Design and Regulation During the Transition to Low-carbon Power Systems, IEA Electricity Market Series, IEA, Paris (2016).

[91] INTERGOVERNMENTAL PANEL ON CLIMATE CHANGE, Climate Change 2013: The Physical Science Basis, Contribution of Working Group I to the Fifth Assessment Report of the Intergovernmental Panel on Climate Change, IPCC, Cambridge University Press, Cambridge (2013).

[92] INTERGOVERNMENTAL PANEL ON CLIMATE CHANGE, Climate Change 2014: Impacts, Adaptation, and Vulnerability, Contribution of Working Group II to the Fifth Assessment Report of the Intergovernmental Panel on Climate Change, IPCC, Cambridge University Press, Cambridge (2014).

[93] INTERNATIONAL ATOMIC ENERGY AGENCY, Adapting the Energy Sector to Climate Change, IAEA, Vienna (2019).

[94] INTERGOVERNMENTAL PANEL ON CLIMATE CHANGE, Managing the Risks of Extreme Events and Disasters to Advance Climate Change Adaptation, IPCC, Cambridge University Press, Cambridge (2012).

Annex I

OPERATING EXPERIENCE FROM MAJOR BLACKOUTS
ON HIGH RELIABILITY GRID SYSTEMS

A number of Member States with normally reliable grid systems have experienced major blackouts (i.e. an event in which electricity supply was lost over a large area) or a system split (i.e. where the action of protection systems following a fault caused an interconnected system to split into two or more parts with large changes in system frequency). For example, in 2003, there were the following well publicized blackouts:

— The United States of America and Canada, 14 August 2003 (see Ref. [I–1]);
— Sweden and Denmark, 23 September 2003 (see Ref. [I–2]);
— Italy, 28 September 2003 (see Ref. [I–3]).

The US–Canada Task Force analysed seven major blackouts in the United States of America that occurred between 1965 and 1999, in addition to the 2003 blackout, and reported their findings in Ref. [I–1]. Some of these blackouts were initiated by severe weather events (e.g. several lightning strikes, collapse of an overhead line in high winds). However, the task force concluded that all major blackouts had a number of common factors that caused the event to grow from a limited grid disturbance to a major blackout, which were not related to the nature of the initiating events but were more to do with the way that the system was operated and managed. These were as follows:

— Overestimation of the dynamic reactive output of generating units (i.e. of the ability of generating units to control system voltage);
— Inability of system operators or coordinators in each control area to visualize events on the entire system (i.e. outside their control area);
— Failure to ensure that system operation was within safe limits (e.g. because of inaccurate modelling or no reassessment of system conditions following the loss of a circuit);
— Lack of coordination on system protection (which resulted in failure to operate or incorrect operation of one or more relays as an event developed);
— Ineffective communication, particularly between different control centres;
— Lack of 'safety nets' (e.g. automatic load shedding or tripping of generation);
— Inadequate training of operating personnel, particularly practice for emergency situations.

Reference [I–1] contains a large number of detailed recommendations related to the design and management of the transmission networks in North America to reduce the risk of similar large cascading events in the future. The analysis of blackouts and system splits in Europe led to similar recommendations, as described in Refs [I–2, I–3].

REFERENCES TO ANNEX I

[I–1]   US–CANADA POWER SYSTEM OUTAGE TASK FORCE, Final Report on the August 14, 2003 Blackout in the United States and Canada: Causes and Recommendations, Department of Energy, Washington, DC (2004).

[I–2]   CLOD-SVENSSON, H.H., Power Failure in Eastern Denmark and Sweden (September 23, 2003), e+i **121** (2004) 367–369.

[I–3]   UNION FOR THE CO-ORDINATION OF TRANSMISSION OF ELECTRICITY, Final Report of the Investigation Committee on the 28 September 2003 Blackout in Italy, UCTE Report, Brussels (2004).

Annex II

CASE STUDY OF QUALITY ASSURANCE IN THE OPERATION AND MAINTENANCE OF SUBSTATIONS AND POWER PLANTS IMPORTANT FOR ELECTRICAL GRID SECURITY

In the late 1980s and early 1990s, the following significant incidents occurred in the French grid system:

— Large brownout in West France tripping eight nuclear power plant units due to the trip of a fossil fuel fired plant, combined with high demand and labour strikes (12 January 1987);
— Simultaneous trips of three 900 MW(e) nuclear power units due to locked protections (1 March 1986);
— Massive load rejection of 1400 MW(e) due to a human error when cabling a cabinet (30 June 1986);
— Loss of 2700 MW(e) generation and loss of off-site power of three 900 MW(e) nuclear power units due to a malfunction of busbar protection after the explosion of a protection current transformer (25 January 1987);
— Tripping of 1400 MW(e) of hydroelectricity generation simultaneously due to a highly resistive fault on a 400 kV line (5 September 1988);
— A single power substation near a nuclear power plant consisting of four 900 MW(e) units experienced two major incidents within six months, one due to a faulty switch and one caused by a fire in a relay cabinet, requiring several nuclear power plants to shut down for several days from 17 November 1991 (the first incident occurred on 30 May 1991).

These incidents led Électricité de France to set up an extensive review and revamp of its quality assurance programme and to place all risky activities under the ISO 9001 standard [II–1] for the security of the electricity system. To accomplish this, two main projects were launched:

(a) The Paramêtres Importants pour la Sureté du Système Electrique (PIS) project:
    (i) To identify all physical parameters in high voltage grid substations and in all large generating units that have a strategic role for system security;
    (ii) To define for each parameter its value, ownership, testing methods and maintenance procedures.
(b) The Maintenance des Postes 400 kV (M400) project:
    (i) To define all critical activities when maintaining a 400 kV substation (including civil works, apparatus installation, electrical cabling and software configuration management);
    (ii) To elaborate maintenance procedures and operation processes that ensure compliance with the ISO 9001 standard [II–1].

The PIS project scope was primarily applicable to the following functional parameters of non-Class 1E equipment:

— Generator and transformer protections;
— Auxiliary protections;
— Voltage control;
— Turbine control;
— Turbine protections;
— Steam control and protections.

These six functional areas cover about 100 physical parameters (e.g. gains, thresholds, delays). Furthermore, each area is associated with a defined function, value and tolerance, as well as the testing method.

The benefits of these two projects have been highlighted by a significant reduction in the number of grid system related incidents since 2000. These projects have allowed the development of a beneficial grid security culture for both nuclear power plant operators (who were focused on generating bulk power) and grid system operators (who were more concerned with minimizing power outages for clients).

REFERENCE TO ANNEX II

[II–1]   INTERNATIONAL ORGANISATION FOR STANDARDIZATION, Quality Management Systems — Requirements, ISO 9001:2015, ISO, Geneva (2015).

Annex III

OPERATING EXPERIENCE FROM LARGE FREQUENCY EXCURSIONS

There have been two large frequency excursions in Great Britain in the recent past, in May 2008 and in September 2019 (see Refs [III–1, III–2]). In both cases, the frequency fell low enough to initiate the first stage of low frequency demand disconnection (LFDD), which successfully prevented the frequency from falling further, and it was possible to restore the system frequency to the normal range within a few minutes. There were no adverse effects on the nuclear power plants, which helped to stabilize the system by remaining at full power throughout.

Great Britain (i.e. England, Wales and Scotland) is a single synchronous area with a single grid system operator (now called National Energy System Operator, previously National Grid ESO), which is required to operate the system in accordance with published standards (see Ref. [III–3]). There are connections to several other networks (including France, the Netherlands, Belgium and Ireland), but these are high voltage direct current (HVDC) and not synchronous connections. The grid system operator normally controls frequency within the range 49.8–50.2 Hz and schedules sufficient frequency containment reserve and frequency restoration reserve to control the frequency fall adequately after the largest single contingency on the day (i.e. the trip of the largest nuclear power unit, or the largest generation loss from a single event on the transmission system), which is typically around 1200 MW. It is acceptable for the system frequency to fall transiently below 49.5 Hz, provided that it can be restored to above 49.5 Hz within 5 min, and such events can be considered 'infrequent'.

If two or more large generation sources are disconnected close together in time, then the scheduled reserves may not be sufficient to control the fall in frequency. The frequency may continue to fall to 48.8 Hz, triggering the first stage in the national LFDD scheme, where up to 5% of the demand would be automatically disconnected by the action of low frequency relays. That is what happened in the two events described here.

In the event of May 2008, described in Ref. [III–1], two large generation sources disconnected independently less than two minutes apart, at a time when demand was rising. This also caused a significant number of small generating units to disconnect, because of the action of rate of change of frequency (ROCOF) relays, or because their low frequency trip setting was set at 49.0 Hz. When the frequency reached 48.8 Hz, 546 MW of the demand disconnected automatically under the LFDD scheme, and the frequency began to rise. The frequency was restored to normal values within 11 min of the initiating event.

The event of September 2019, described in Ref. [III–2], started with a lightning strike at a time of rising demand. This caused the steam turbine, and later the gas turbines, at a nearby combined cycle gas turbine power plant to trip and caused a more distant large off-shore wind farm to shut down. In addition, a substantial number of small generators fitted with vector shift relays or ROCOF relays disconnected. Some more small generating units disconnected on low frequency protection, set at 49.0 Hz. When the frequency reached 48.8 Hz, the LFDD scheme disconnected 931 MW of the demand, but it also disconnected around 581 MW of the embedded generation, so the net reduction of demand was only 350 MW. Nevertheless, the frequency was restored to normal values within 5 min of the initiating event.

The first event was caused by the coincident but independent loss of two generating sources, although the frequency transient was made worse by the disconnection of a large number of small generating units as a result of the frequency transient. The second event resulted from two large generation sources being sensitive to a voltage transient, but was made more significant because a large number of small generating units were also sensitive to the voltage and frequency transient. The technical rules for large generating units (see Ref. [III–4]) require the generating units to have fault ride-through capability, so the large generating units were not expected to disconnect in these circumstances. The technical rules for small generating units (see Ref. [III–5]) now also require that they are similarly robust against voltage and frequency transients. However, these requirements have come into force only recently for new small generating units, and a large number of existing small generation sources do not have this robustness, and it is difficult to impose new requirements on existing units.

One learning point for Member States is that to minimize frequency transients, all generating units need to be similarly robust against voltage and frequency transients. It is important to have suitable technical requirements in force for small generating units before many such units are connected. Another learning point is that an LFDD scheme can become less effective if there is a substantial number of small generators connected at low voltage within distribution networks, which can become simultaneously disconnected with demand.

REFERENCES TO ANNEX III

[III–1]   NATIONAL GRID ESO, Report of the National Grid Investigation into the Frequency Deviation and Automatic Demand Disconnection that Occurred on 27th May 2008, National Grid, Warwick (2009).

[III–2]   NATIONAL GRID ESO, Technical Report on the events of 9 August 2019, National Grid, Warwick (2019).

[III–3]   NATIONAL GRID ESO, Security and Quality of Supply Standard (SQSS), Ver. 2.4, National Grid, Warwick (2019).

[III–4]   NATIONAL GRID ESO, The Grid Code, Issue 6, Rev. 20, National Grid, Warwick (2023).

[III–5]   ENERGY NETWORKS ASSOCIATION, The Distribution Code of Licensed Distribution Network Operators of Great Britain, Issue 55, ENA, London (2023).

EXAMPLE OF MARKET ACTIVITIES AFFECTING GRID PLANNING

Deregulation of the electrical systems and changes in electricity markets have modified power exchanges between generators and electricity consumers and have had consequences for the electrical grid systems in some Member States. These consequences have required various changes to grid operation and planning. Fairly steady generation patterns that had been used in designing and planning transmission networks in the past have become far less predictable and more variable.

These changes have affected grid reliability because of the increased difficulty of planning adequate reserves of active and reactive power. Figure IV–1 provides a simplified example of how market activities can cause grid operational and planning uncertainties.

In Fig. IV–1, there are two generators, A and B, producing electricity for €30/MW·h and €32/MW·h, respectively, economically dispatched to 600 MW of generator A and 300 MW of generator B, since the total load is 900 MW. Since the transmission line ratings are 300 MV·A for all five lines, the load flows in the network are such that the line between nodes 1 and 2 will be overloaded, so the grid system operator may consider reinforcing this line by constructing a parallel one. If the grid system operator invests in the second line, but generator B changes its bid in the future down to €29/MW·h (or generator A increases its bid up to €33/MW h) because of new market conditions, the line between nodes 3 and 4 will be overloaded and network reliability will still be jeopardized despite the previous grid reinforcement. In this example, an alternative remedy might have been to install one or two phase shifting transformers to control power flows.

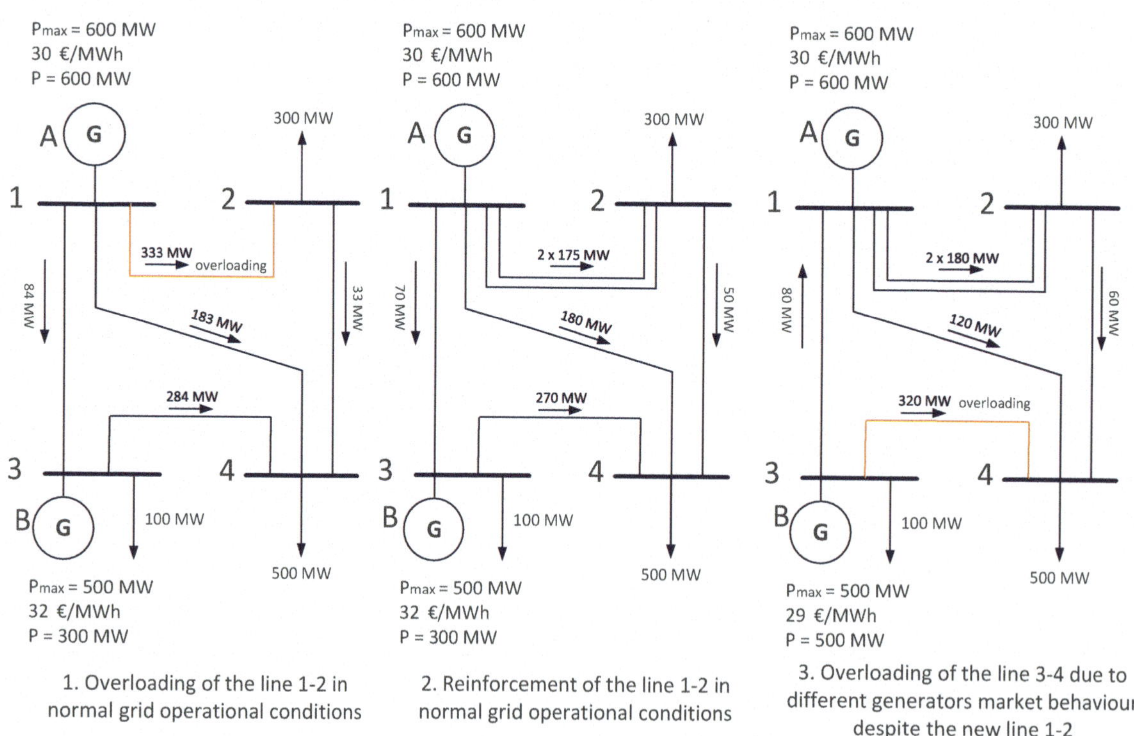

FIG. IV–1. Example of deteriorating grid reliability due to market uncertainties. Reproduced courtesy of D. Bajs, Energy Institute Hrvoje Pozar.

GLOSSARY

active power. The product of the magnitude of the voltage and current and the cosine of the phase angle between them, normally expressed in megawatts (MW). It is sometimes called real power or true power.

ancillary service. A service that may be provided by generating units to assist the control of the grid system. Ancillary services can include provision of reactive power and control of system voltage, automatic frequency and generation control, load following, provision of spinning reserve, standing reserve and black start capabilities and emergency control actions. In electrical systems that have been deregulated, there would normally be a payment to generators for providing ancillary services and a requirement in the grid code for them to have the necessary capability.

apparent power. The product of the magnitude of the voltage and current, ignoring the phase angle between them, usually expressed in megavolt amperes (MV·A). The full load rating of generators and transformers is normally expressed in MV·A.

automatic frequency control. A method of operating a generating unit at less than full output, under automatic control, so that its output increases automatically if the system frequency falls and decreases automatically if the system frequency rises.

automatic generation control. A method of operating a generating unit so that its output can be changed automatically in response to a signal from the grid control centre, without requiring actions by the operators at the power plant.

auxiliary equipment. See 'plant auxiliary equipment'.

auxiliary transformer. A transformer that provides electrical supply from the generator to one or more buses that supply the main electrical auxiliary equipment of the reactor. Some nuclear power units have more than one auxiliary transformer. In most nuclear power plant designs, the auxiliary transformer(s) provide most or all of the power to the auxiliary electrical equipment when the reactor is on load. In other publications, the auxiliary transformer may be called a unit transformer or unit auxiliary transformer.

balancing services. All actions and processes, on all timescales, through which grid system operators maintain the system frequency within a predefined range. Balancing services are generally provided by controllable generating units, but additional services may be provided by controllable demand. Balancing services are a subset of ancillary services.

blackout. A condition in which all electrical power has been lost in all or a large part of the grid system in the country or region. To be distinguished from 'station blackout'.

black start capability. The capability of a power plant to start generation from a shutdown condition in a short time, with no external source of power. A sufficient number of power plants or other power sources with this capability are essential to restore normal operation of a power system after a widespread blackout.

brownout. A condition in which the voltage on a part of the electrical grid falls to a low level following a fault, and the voltage control arrangements are not sufficient to restore the voltage, so the voltage continues to fall for one or two minutes. If the network operator does not start rapid remedial action, the voltage will continue to fall and lead to a blackout.

capacity market. Commercial arrangements to ensure sufficient generation capacity is always available. This can include payments to power plants for their available generating capacity, independently of payments for energy generated. It can also include payments to electricity consumers with large demand who agree to reduce demand when instructed.

central dispatch. A system used in some Member States where the grid control centre decides the planned generation profile for each generating unit, based on prices or system needs. Generating units are expected to follow this planned profile unless later instructed otherwise by the grid control centre.

degraded power. Conditions of off-site power supply that risk preventing a nuclear power plant system or component from fulfilling its intended function for safety and performance. This could include prolonged high or low voltage or abnormal frequency or unbalanced phases.

dependability. Typically, the combination of availability (i.e. readiness for correct service), reliability (i.e. continuity of correct service), security (i.e. absence of catastrophic consequences for the users and the environment), integrity (i.e. absence of improper system alteration) and maintainability (i.e. ability for a process or equipment to undergo modifications and repairs).

dispatch. The issuing of instructions or signals by the grid control centre to generating units to increase or decrease electrical output.

dispatchable generation. Power generation by a generating unit that is able to receive and act on instructions or signals from the grid control centre to adjust its output. As grid control centres do not generally have a means of communicating with small generating units that are connected to distribution networks, such small units are non-dispatchable.

distribution system. Parts of the public electrical system that are operated at lower voltages. (The typical voltages are 33 kV, 11 kV, 400/230 V, but this varies between countries.) The distribution system connects the transmission system to electricity consumers supplied at low voltage. The configuration and behaviour of the distribution systems are not generally relevant to nuclear power plants, which are normally connected only to the transmission system.

droop. The reciprocal of the gain of the frequency control loop of the governor on a generating unit. A 4% droop means that a 4% change in frequency would cause a 100% change in output. The droop is commonly set in the range 3–5%.

electrical grid. For this publication, the public electrical system to which the nuclear power station and power plants of all kinds are connected, which comprises a transmission system and distribution systems.

energy only electricity market. A wholesale electricity market where generating units are paid only for the energy that they export (in MW·h) and are not paid for capacity or for ancillary services.

fault. For this publication, any unplanned event that causes a circuit or an item of equipment in a network to be switched out of service, either automatically by the electrical protection system or manually in response to an alarm signal.

fault ride-through capability. The ability of a generator to maintain operation during and after a transient low voltage event, such as one caused by an electrical fault close to the generator.

frequency response. The magnitude of the change in generator electrical output power that can be achieved within a certain time in response to a change in system frequency.

generator breaker. A circuit breaker fitted at the terminals of a generator. Not all nuclear power plants have such a breaker fitted. If it is fitted, then opening the breaker would allow the plant auxiliary equipment to be supplied from the grid system via the unit and auxiliary transformers when the generator is not in operation.

grid code. A set of technical requirements that grid system operators, power plant operators and sometimes large consumers must follow to ensure the safe, reliable and efficient operation of an electrical grid. It typically covers areas such as voltage control, frequency stability, protection systems and operational procedures.

grid control centre. The central operation structure and organization that contains equipment and personnel that provide high level control of the grid system in a country or region. It issues instructions to generating units to increase or decrease output, or to start up or shut down, in order to balance power generation with load. It also issues instructions to operate in automatic frequency control mode and to control voltage. The grid control centre may also control the transmission system, switching circuits into or out of service as necessary. A large grid system may have several grid control centres.

grid system operator. For this publication, the company or organization that is responsible for the security of the electrical grid in the area under its control and operates the grid control centre. In some Member States, the grid system operator is the same company that owns and maintains the high voltage transmission system (transmission system owner), while in some other Member States, the grid system operator is a separate organization, termed an 'independent system operator' or 'transmission system operator'.

house load. The power consumption of all the electrical auxiliary plant in the power plant.

house load operation. The ability of a nuclear reactor to remain at power (i.e. reactor critical) if there is an unplanned event that disconnects the nuclear power plant from the electrical grid. The nuclear power unit continues to provide electrical power to the house load, and the reactor control system rapidly reduces the reactor power to allow stable operation in this low power condition.

inertia constant. A measure of the inertia of a synchronous generating unit that contributes to the total system inertia. It is the stored kinetic energy of rotation of the rotor of the generating unit at normal rotational speed (in MW·s) divided by the machine rating (in MV·A), and has units of seconds.

island mode operation. The operation of a generating unit disconnected from the main transmission system, but still connected to part of the transmission system near the power plant with some connected demand. This is similar to house load operation, but with additional demand external to the power plant.

load following. Varying the output of a generating unit in a planned way, or in response to an instruction or control signal from the grid control centre, reducing the output when the load (electrical demand) on the system is reduced (e.g. at night, at weekends, on public holidays), and increasing the output to maximum when the electrical demand is high.

load shedding. A rapid reduction in the electrical output of a generating unit, which could be by tripping the unit or by steam bypassing the turbine. Also called 'power cutback'.

loss of off-site power. The simultaneous loss of preferred off-site AC power to all safety buses (LOOP). (Note: it is likely that the non-safety buses will also lose power.) LOOP may result from events totally within the nuclear power plant or from events external to the nuclear power plant.

low frequency demand disconnection. A protection scheme, used in most countries, where some electricity consumer demand is automatically disconnected in extreme events where the frequency falls to a low value, in order to halt the fall in frequency and avoid a widespread system collapse and blackout.

$N-k$ **criterion.** A criterion for proper operation of a transmission system consisting of N elements, with k elements out of operation or inoperable. For example, if the design and operation criterion is that any one (1) element in an N element system is not functioning, then it is called an $N-1$ criterion.

nuclear power plant operating organization. The company or organization operating a nuclear power plant. This organization has the primary responsibility for the safe operation of the plant and will have to satisfy the requirements of the nuclear regulatory body in the country. The operating organization may not be the owner of the plant.

nuclear power unit/nuclear power generating unit. Comprises a nuclear reactor and all the auxiliary equipment (e.g. generator, transformers, motors, pumps, electrical supplies, protection systems) that are required for its operation. A nuclear power plant may have one or more nuclear power units.

off-load tap changer. A transformer tap changer that may only be operated when the transformer is isolated (i.e. has no voltage applied to any windings). Operating an off-load tap changer when the transformer is energized, even if it is not carrying any load, is almost certain to cause a dangerous catastrophic failure.

off-site power system. The system that normally provides AC power to the nuclear power plant in all modes of operation and in all plant states. It also provides transmission lines for outgoing power.

on-load tap changer. A transformer tap changer that may be operated while the transformer is energized, whether or not the transformer is carrying load.

plant auxiliary equipment. The equipment required for the operation of a nuclear power unit (e.g. transformers, motors, pumps, battery chargers, electrical supplies, protection systems), most of which requires electrical supply for operation.

point of common coupling. The point in an electrical system where a power plant or electricity consumer is connected to the grid system. For a nuclear power plant, this is usually a busbar in the substation near the power plant.

pole slip. An event where a synchronous generator loses synchronism with the local grid voltage. A pole slip can cause significant mechanical damage to a generator if it is not rapidly disconnected from the grid.

power cutback. See 'load shedding'.

power quality. Good power quality is where the sinusoidal voltage of the electricity supply is within a narrow range of frequency and voltage, the voltage of the three phases is balanced, the waveform of each phase is closely sinusoidal with a low level of harmonic distortion and few transient spikes or dips, and the amplitude of the voltage is steady with little flicker.

power transfer capability. In a nuclear power plant, the ability to rapidly transfer the electrical supply to plant auxiliaries from the generator terminals via the auxiliary transformer(s) to the external electrical grid via the standby transformer(s).

preferred power supply. In a nuclear power plant, the power supply from the transmission system up to the safety classified electrical power system. Some portions of the preferred power supply are not part of the safety classification scheme.

quasi-steady state frequency. A measured value of frequency at a time when it is changing slowly enough that it can be considered to be constant.

rate of change of frequency relay. A protection relay that will disconnect a power generation source if the rate of change of frequency (Hz/s) exceeds a preset value. Such relays are commonly used within distribution systems as a means of disconnecting generation sources if there is a loss of connection to the main transmission system. They can cause spurious tripping if there is a rapid fall of system frequency after the loss of a large generating unit.

reactive power. Product of the magnitude of the voltage and current and the sine of the phase angle between them, normally expressed in units of megavars (megavolt-amperes reactive). The reactive power can be positive or negative. The inductance of overhead lines and of the motor load within consumer demand is said to absorb reactive power, while the capacitance of underground cables is said to produce reactive power. Generating units and reactive compensation devices need to produce or absorb varying amounts of reactive power as necessary to balance this, which is important for controlling the voltage on the transmission system.

reserve. Comprises the additional generating capacity that is available to the grid system operator within defined intervals of time. Also known as operating reserve. It includes the frequency containment reserve and the frequency restoration reserve, which are available within seconds or minutes, and replacement reserves, which are available on longer timescales. In some systems, controllable demand also contributes to operating reserves.

rotor angle. In a synchronous machine, the angle between the field magnetic flux and the resultant air gap flux. Changes in the rotor angle will cause changes in the electrical power output. Also known as load angle or power angle.

self-dispatch. A system used in some Member States where the operators of a generating unit declare their planned generation profile to the grid control centre, and then operate the generating unit to match this profile, unless instructed by the grid control centre to do otherwise.

spinning reserves. Additional generation that is available from generating units operating at less than full load that can rapidly increase or decrease output, when necessary, either when instructed by the grid control centre, or automatically for units operating in automatic frequency control mode.

standby transformer. For this publication, a transformer connected to a grid substation that can provide supply directly to the electrical auxiliaries in the nuclear power plant. The high voltage side of the transformer may be connected to the same substation at the same grid voltage as the unit transformer, or it may be connected to a different substation at a lower voltage. The low voltage side of the standby transformer supplies the highest auxiliary voltage used in the nuclear power plant (typical values are 13.8 kV or 11 kV, but this varies between countries). The standby transformer provides the off-site electrical supply to the auxiliaries when the supply via the unit transformer and auxiliary transformer is not available. (In other publications, this transformer may be called the station transformer, startup transformer or backup transformer.)

standing reserves. Additional generation available from generating units that are in a shutdown condition, but are able to start up rapidly, when necessary, either automatically or when instructed by the grid control centre.

station blackout. A plant condition with the complete loss of all AC power from the preferred off-site AC sources, from the main generator and from standby AC power sources important to safety, to the essential and nonessential switchgear buses. DC power supplies and uninterruptible AC power supplies derived from them may be available as long as batteries can supply the loads. Alternative AC power supplies may be available.

station blackout coping period. The specified duration of station blackout for which a nuclear power plant is able to cool the reactor core and maintain containment integrity, and then recover.

substation. Also called 'switchyard'. An installation on the transmission system to which transmission circuits, generating units and consumer load may be connected, and which has facilities for switching. Equipment in a substation can include circuit breakers and other switches, bus bars, transformers and protection and control equipment. A nuclear power plant would normally be connected to one or two substations located close to the nuclear power plant site.

switchyard. See 'substation'.

synchronizing torque. The torque that acts on the shaft of a synchronous machine when the rotational speed of the rotor deviates from the synchronous speed and that keeps the machine in synchronism.

system frequency. The frequency of the alternating voltage on the system. In an interconnected system, the frequency is the same throughout the system at any instant in time. The nominal frequency in almost all countries is either 50 Hz or 60 Hz.

tap changer. A mechanical switching device in a transformer, which is used to change the output voltage of the transformer by altering the number of turns in one winding and thereby changing the turns ratio of the transformer. There can be an on-load tap changer or an off-load tap changer.

transmission system. Those parts of the public electrical system operated at high voltages (typical voltages are 400 kV, 275 kV, etc., but this varies between countries). The transmission system is used to interconnect large power stations with centres of load and to transmit large amounts of power long distances. Nuclear power plants are normally connected to the transmission system, so the design and performance of the transmission system is important to nuclear power plants' safety and performance.

transmission system owner/operator. The company or organization that owns and maintains part or all of the transmission system. In some Member States, this is the same organization as the grid system operator, in others it is independent. In some Member States, there is a single transmission system owner for the whole country, in others, there are several transmission system owners.

unit transformer. For this publication, the transformer connected between the generator (alternator) and the transmission system. The unit transformer is the route for export of generated power from the nuclear power plant. (In other publications this may be called the generator transformer, the block transformer or step up transformer.)

ABBREVIATIONS

AC	alternating current
ANSI	American National Standards Institute
CIGRE	International Council on Large Electric Systems
DC	direct current
EDG	emergency diesel generator
HVDC	high voltage direct current
IEC	International Electrotechnical Commission
LFDD	low frequency demand disconnection
LOOP	loss of off-site power
NERC	North American Electric Reliability Corporation
OLCs	operational limits and conditions
ROCOF	rate of change of frequency
SBO	station blackout
SSCs	structures, systems and components
WANO	World Association of Nuclear Operators

CONTRIBUTORS TO DRAFTING AND REVIEW

Alzieu, M.	Électricité de France, France
Bajs, D.	Energy Community, Croatia
Del Rosso, A.	Electric Power Research Institute, United States of America
Duchac, A.	Nuclear Research Institute Řež, Czech Republic
Grgic, D.	University of Zagreb, Croatia
Kerr, C.	Electric Power Research Institute, United States of America
Kilic, N.	International Atomic Energy Agency
Leake, H.	Consultant, United States of America
Lebreton, P.	Consultant, France
Lundbäck, M.	Svenska kraftnät, Sweden
Marquet, J.N.	Électricité de France, France
VanHeek, A.	Nuclear-21, Austria
Volkanovski, A.	Jožef Stefan Institute, Slovenia
Ward, D.M.	Consultant, United Kingdom
Yu, Q.	International Atomic Energy Agency

Technical Committee Meetings

Zagreb, Croatia: 20–22 September 2016
Stockholm, Sweden: 15–17 October 2019

Consultants Meetings

Vienna, Austria: 9–12 May 2016; 30 October–1 November 2017;
19–21 June 2018; 15–18 October 2019;
5–8 April 2022

Structure of the IAEA Nuclear Energy Series*

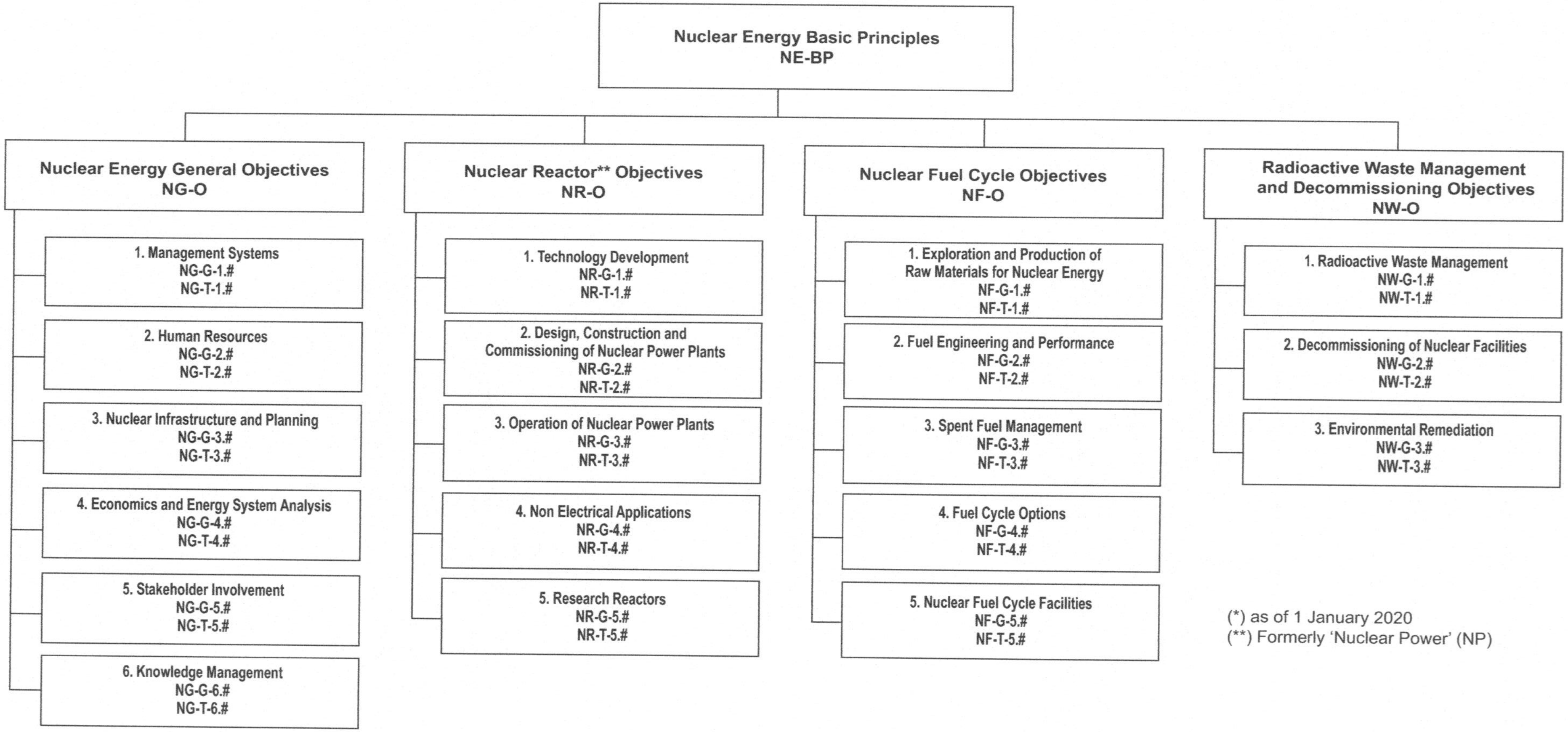

(*) as of 1 January 2020
(**) Formerly 'Nuclear Power' (NP)

Key

BP: Basic Principles
O: Objectives
G: Guides and Methodologies
T: Technical Reports
Nos 1–6: Topic designations
#: Guide or Report number

Examples

NG-G-3.1: Nuclear Energy General (**NG**), Guides and Methodologies (**G**), Nuclear Infrastructure and Planning (topic **3**), **#1**
NR-T-5.4: Nuclear Reactors (**NR**), Technical Report (**T**), Research Reactors (topic **5**), **#4**
NF-T-3.6: Nuclear Fuel (**NF**), Technical Report (**T**), Spent Fuel Management (topic **3**), **#6**
NW-G-1.1: Radioactive Waste Management and Decommissioning (**NW**), Guides and Methodologies (**G**), Radioactive Waste Management (topic **1**) **#1**

CONTACT IAEA PUBLISHING

Feedback on IAEA publications may be given via the on-line form available at:
www.iaea.org/publications/feedback

This form may also be used to report safety issues or environmental queries concerning IAEA publications.

Alternatively, contact IAEA Publishing:

Publishing Section
International Atomic Energy Agency
Vienna International Centre, PO Box 100, 1400 Vienna, Austria
Telephone: +43 1 2600 22529 or 22530
Email: sales.publications@iaea.org
www.iaea.org/publications

Priced and unpriced IAEA publications may be ordered directly from the IAEA.

ORDERING LOCALLY

Priced IAEA publications may be purchased from regional distributors and from major local booksellers.

Printed and bound by CPI Group (UK) Ltd, Croydon, CR0 4YY

06/07/2026

02160600-0008